PHYSIOLOGICAL APPROACH
TO THE LOWER ANIMALS

J. A. RAMSAY

*Professor of Comparative Physiology
and Fellow of Queens' College in the
University of Cambridge*

SECOND EDITION

CAMBRIDGE
AT THE UNIVERSITY PRESS
1972

CAMBRIDGE UNIVERSITY PRESS
Cambridge, New York, Melbourne, Madrid, Cape Town,
Singapore, São Paulo, Delhi, Tokyo, Mexico City

Cambridge University Press
The Edinburgh Building, Cambridge CB2 8RU, UK

Published in the United States of America by Cambridge University Press, New York

www.cambridge.org
Information on this title: www.cambridge.org/9780521095372

First edition 1952
Reprinted 1958, 1962, 1964
Second edition 1968
Reprinted 1972
Re-issued 2011

A catalogue record for this publication is available from the British Library

Library of Congress Catalogue Card Number: 68–21398

ISBN 978-0-521-09537-2 Paperback

CONTENTS

PREFACE TO THE FIRST EDITION

One of the most notable features of zoology courses for the more advanced classes in schools is the relatively large number of periods devoted to the structure and physiology of the mammal. This is probably not peculiar to Great Britain, but the syllabus drawn up by the Cambridge Joint Advisory Committee for Biology and published in 1944 may serve as a representative example of an attitude that seems universal and almost inevitable. In making these proposals the Committee state: 'It is our intention that the teacher should accept the student's natural anthropocentric view as a starting point. He can then show that the special problems of our own bodies are in fact only part of the general problem of the nature of living matter and that man himself finds a natural place among a variety of other living organisms.' The Committee propose that the course should begin with 'The human body, its structure and function', and their next section is 'The variety of animals'. The Committee do not specifically mention physiology in this section; but, as may reasonably be inferred from the passage quoted above, they no doubt had in mind that opportunities would present themselves for drawing comparisons between the functional systems of the invertebrates and those of the mammal previously studied.

While being wholeheartedly in favour of taking the human body as the starting point I must admit that I find the next step a little awkward. Are we to take *Hydra* next, only to confess that we know almost nothing about its physiology and that human physiology does not help? Or are we to work downwards from the mammals to the lower animals and derive our general principles at the end? This latter line of approach may be a logical one but it is difficult for the student; like the reader of a detective story he is kept guessing until the last chapter. In my experience the student finds it very much easier to be presented with the general principles at an early stage and to have their application illustrated later. Taking this line, the problems of digestion, respiration and so on may be set forth once and for all at the biochemical level; but at the physiological level they present themselves in quite different ways according to the size of the animal, the plan upon which its body is organized and its mode of life. It is this particular physiological approach to the lower animals that I have in mind. I would like the student to realize that the mammal is a

v

highly-tuned physiological machine carrying out with superlative efficiency what the lower animals are content to muddle through with, and that he must not necessarily expect to find in the lower animals the physiological processes and their special organs which he found in the mammal.

I have deliberately restricted myself to such parts of the subject as are susceptible to broad generalization; but the broader the generalization the more exceptions it has to admit of. In disregarding such exceptions I realize that I have laid myself open to the charge of over-simplification. Various friends who have been kind enough to read this book in typescript have been at pains to call my attention to this and that unguarded statement, to other interpretations which have been put upon the facts I have quoted, and so on. To most of these criticisms I have been able to reply that the omissions have been deliberate. I cannot see much point in stating a generalization only to smother it with qualifications. By the time a student comes to the university he is already sadly aware that it is in the nature of rules to have exceptions and he is not going to be utterly disillusioned when in due course exceptions come to his notice. Why distract him with them at this stage?

As I understand it, a text-book is written to cover a certain field of knowledge up to a certain standard of knowledge, and it is the task of the author of a text-book to seek a balanced treatment of his subject, allotting to each aspect such space as its importance seems to demand. If this be so, let me state at once that this volume is not intended to be a text-book of comparative physiology. It is intended to be an introduction to the subject, a discussion of certain principles and their application but not in any sense a survey of invertebrate physiology. I have therefore felt at liberty to leave out parts of the subject which do not lend themselves readily to my purpose. I have said very little about hormones and nothing at all about reproduction, or about chromatophores and independent effectors. I have not prepared an index because I cannot imagine that anyone will want to use this book as a work of reference.

The book is intended for the student at the beginning of his university career but I hope that it may also be of use in schools at the scholarship level. I have assumed that the reader has worked through a syllabus similar to that advised by the Cambridge Committee—I understand that it is still approved in England for the more advanced school-leaving examinations—and without submitting to any rigid restriction I have tried to find illustrations and examples from among the animals specified therein. In an Appendix I have given a brief classification of the Animal Kingdom and this is more detailed than that in the Syllabus only in respect of the three classes of Mollusca. I have also assumed basic knowledge of physics and chemistry.

I wish to take this opportunity of thanking Professor J. Gray, Dr G. Salt, Dr G. S. Carter, Dr J. W. S. Pringle, Dr J. W. L. Beament and Dr G. A. Kerkut for reading parts of this book in typescript and for their advice and criticism. I also wish to thank Dr C. F. A. Pantin, Dr V. B. Wigglesworth and Professor E. H. F. Baldwin for permission to reproduce certain figures from their publications.

Queens' College
Cambridge

J.A.R

PREFACE TO THE SECOND EDITION

When I undertook to prepare this second edition the first thing I did was to look at the date of the first—1952. It was something of a shock to realize that this was the year before Watson and Crick published the structure of DNA. I can even remember deciding not to say anything about mitochondria or ATP on the grounds that their general significance was not yet firmly established. What changes have we not seen in these fifteen years!

Reading over what I had written in 1952, I was agreeably surprised to find so little that I wished to withdraw. This is of course as it should be if one has sought to present only the principles of a subject at an elementary level. Yet it is a matter of some astonishment to me that the sales of this book have been so long maintained, for the presentation of these principles is now hopelessly outdated. In 1952 I was writing for a public brought up in traditonal zoology, with its emphasis upon the types of animal organization. It was even perhaps a little *avant garde* in those days to suggest that comparative physiology should be introduced so early in the student's experience. But all that is changed now. Thanks to the popularization of science, encouraged by journals like the *Scientific American* and the *New Scientist*, the student's interest is already awakened, even at school, in the latest developments. And the revolution which has taken place in biology as a science has also taken place in biology as a school subject. The separate treatment of zoology and botany is now being replaced by a unified treatment of biology at the cellular level, in which biochemistry figures prominently.

To conform with this changing background of knowledge I have had to recast many of the chapters. I have put cellular processes to the fore under respiration, though eschewing the biochemical details which are too often allowed to clutter up this subject. I have introduced the sliding filament theory of muscular contraction. I have even touched upon the ionic theory of the nerve impulse and tried to relate it to ionic regulation in cells. I like to think that these changes have made for a more logical presentation without greatly adding to the length of the book, though they have added to the number of chapters. The only chapter which is entirely new is Chapter 5, on chemical coordination, which I have for long felt should not have been omitted from the first edition. I am grateful to Sir Vincent

Wigglesworth for reading this chapter in draft and for his most helpful comments. Many of the diagrams have been re-drawn with
assistance from the Cambridge University Press, whose cooperation
in this and in so many other matters it is a pleasure to acknowledge.

Some of the reviewers, whose comments upon the first edition I
had found encouraging, were disposed to note that the book was
without an index and without any references for further reading. My
views about the index remain as they were—that I cannot imagine
anyone wanting to use this book as a work of reference. Nevertheless
I have prepared an index and it is, in my opinion, quite useless. On
the matter of further reading I recognize that I have been at fault, but
I have also been in some difficulty. It is not to be forgotten that the
student of today has little enough time to read around his subjects,
and suggestions as to further reading must be made with this in mind.
Has one not too often seen that deflated look upon the face of a
student who has brightly asked an innocent question and now holds
in his hand the list of references which an over-zealous instructor has
advised him to consult. I made up my mind that to list some works
for further reading, baldly and without comment, at the end of each
chapter—as is so often done—would not meet the case. Instead, I
have added an Appendix which is intended to help the student to find
his own way into the more advanced parts of the subject and to be
discriminating in his choice of reading matter as he does so.

Queens' College   J.A.R.
Cambridge

September 1967

1

NUTRITION

Animals require food mainly for two reasons: first, for the building up of the substance of their own bodies, which we will call the *synthetic requirement*, and second, for the provision of energy, which we will call the *energy requirement*. We will deal with the energy requirement first.

There are three main categories of energy requirement which we can recognize in animals:

1. Energy for work: this is obvious, but one should remember that the muscles are not the only organs of the body which do work; the formation of secretion by a gland nearly always involves work in the osmotic sense.

2. Energy for synthesis: the complex materials of the body have a higher energy content than the simpler substances out of which they are built.

3. Maintenance energy: this last requires further explanation.

Even when it is at rest and when digestion and other physiological processes are in abeyance an animal continues to make use of energy on a considerable scale. In the case of a warm-blooded animal much of this energy appears directly as the heat required to maintain the temperature of the body. But even in the simplest forms of life, where nothing much seems to be happening, there is still a continuous expenditure of energy and if the supply of energy fails death very soon follows. An animal is not like a motor-car which can be parked at the side of the road with the engine switched off: it must have its engine kept running, it must have petrol continually supplied, or it will simply fall to pieces.

This analogy can usefully be taken a little further. Every so often a motor-car must be taken into a garage to be overhauled and to have its worn-out parts replaced. No such convenient intermission of its activities is possible for an animal. Think of your own heart, contracting regularly about seventy times a minute and likely to keep it up for some seventy years. An animal has to carry out all its repairs on the move.

The repair services of the body are a very active organization and of recent years we have been finding out something more about them

I

by the use of isotopes. To take an example. An animal is fed with an amino-acid in which the nitrogen atoms of the amino-groups are 'heavy'—atomic weight 15 instead of 14. These 'heavy' nitrogen atoms can later be detected in various parts of the body; one molecule of the amino-acid becomes incorporated into the protein of a muscle fibre, another is deaminated and its nitrogen appears in the urine in the form of urea, and so on. What emerges from these studies is that there is a continual turnover of the materials of the body. Enzyme molecules seem to wear out very quickly and have to be replaced by new ones; and not only do we find the body, like a good housewife, turning over its food reserves in the liver and in the fat deposits, but even structures like bones, which we tend to think of as permanent, are brought within the system of renewal and replacement.

All organisms—animals, plants, and bacteria—have in common these requirements for energy, but they satisfy their requirements in a variety of ways. In the animal body energy is furnished by the oxidation of organic substances to water and carbon dioxide. Although plants can use the energy of sunlight to synthesize simple organic compounds they also make use of oxidative mechanisms as do animals; in the absence of sunlight they absorb oxygen and give off carbon dioxide. But in the bacteria we find an astonishing diversity in the means of obtaining energy. There is a whole range of putrefactive bacteria of meat and fermentative bacteria of milk which obtain energy by breaking down organic substances such as are used by animals. *Clostridium botulinum* is one which grows in protein material in the absence of oxygen. It may be present in tinned food which has not been adequately sterilized before sealing and it produces a toxin, probably the most poisonous substance in the world, which causes that form of food poisoning known as botulism. Although it makes use of the same food material as we do, the fact that it thrives in the complete absence of oxygen shows clearly that it must obtain its energy by processes entirely different from our own. *Nitrosomonas*, one of the nitrifying bacteria of the soil, gets its energy from the oxidation of ammonia to nitrite. The sulphur bacteria oxidize sulphur to sulphuric acid. The iron bacteria oxidize ferrous carbonate to ferric hydroxide, and their activities often cause the blocking of iron pipes with rust. There is even a bacterium which can live by oxidizing hydrogen to water. In comparison with bacteria animals seem to show only conservatism and lack of resource in their biochemical methods.

We turn now to the material requirement. The bulk food of animals consists of protein, carbohydrate and fat, all of which are complex

2

organic compounds. In the alimentary canal these substances are broken down by digestive enzymes to compounds of a lower order of complexity, proteins to amino-acids, carbohydrates such as starch to glucose, fats to glycerol and fatty acid, and these simpler compounds are absorbed into the body. One would therefore imagine that it would be possible to keep an animal alive and healthy on a diet of amino-acids, monosaccharides, glycerol and fatty acids—and up to a point this is true. Indeed, one could simplify its diet even further since protein, carbohydrate, and fat are to some extent interconvertible in the body. An animal can be fattened on a diet of carbohydrate and when it is starved it can live off its fat. It has been shown that carbohydrate can be formed from protein. But what the animal cannot do is to make protein out of carbohydrate or fat, even if it has a supply of nitrogen available. Given one kind of protein in its food the animal can break it down to its constituent amino-acids and it can then build up these amino-acids into other proteins. There are about twenty amino-acids in natural proteins and some of these can be synthesized in the body, but—at least in the rat, where this matter has been most closely studied—there are ten essential amino-acids which cannot be synthesized in the body but must be provided in the food. Compare this with what happens in a plant. The material requirements of a plant are simple inorganic compounds, water, carbon dioxide and nitrate. Given these and a supply of energy the plant can synthesize the whole variety of organic compounds which go to build up its tissues. Organisms which can make use of simple inorganic materials are said to be autotrophic. Plants are photosynthetic autotrophes, using the energy of sunlight. Many bacteria are chemosynthetic autotrophes, obtaining energy from chemical reactions such as those mentioned in the last paragraph.

Let us note, therefore, that in comparison with other organisms animals are highly exacting in the material requirements of their diet. Let us note also that, since animals obtain their energy by oxidizing organic compounds such as are used in building up their own bodies, their energy requirements are closely bound up with their material requirements. In the chemosynthetic autotrophes the substances which satisfy the material requirements are not necessarily concerned in the provision of energy.

The essential amino-acids are part of the bulk material requirement and relatively large quantities must be provided in the diet. If the animal is to be kept in health its diet must also contain vitamins, but of these relatively small quantities are sufficient. The word vitamin came into use in the period 1910–20 when it became realized that purified preparations of the bulk constituents of the food were inadequate to maintain health. Since then nearly twenty vitamins have been recognized; many have been obtained in pure form, some have been synthesized in the laboratory, and in a few cases we have discovered their special significance to the body. We will take one example only, that of vitamin B_1, now identified as thiamin, lack of which causes the disease 'beri-beri'. The biochemical role of thiamin was first discovered in yeast. Thiamin pyrophosphate is a part (co-enzyme) of the enzyme carboxylase which is responsible for the breakdown of pyruvic acid, CH_3CO $COOH$, to acetaldehyde and carbon dioxide. In the animal body pyruvic acid is formed during the oxidation of carbohydrate and the further oxidation of pyruvic acid to water and carbon dioxide does not take place in the absence of thiamin. The structural formula of thiamin is given below—and shows how it is formed by the combination of one molecule of pyrimidine with one molecule of thiazole. Neither the thiamin molecule nor its pyrimidine and thiazole moieties can be synthesized in the bodies of the higher animals; but in the flagellate protozoa we see an interesting variation in synthetic ability. Some species have no more synthetic ability than a mammal and will not grow in culture unless thiamin is supplied. Some cannot synthesize pyrimidine and thiazole, but given these two molecules they can combine them to form thiamin. Some can synthesize pyrimidine and combine it with supplied thiazole, others conversely can synthesize thiazole and combine it with supplied pyrimidine. The great majority can carry out the whole synthesis from simple inorganic compounds.

Pyrimidine   Thiazole

Thiamin, vitamin B_1

This general discussion of the nutritional requirements of organisms brings out the fact that the synthetic abilities of the lower organisms appear greatly to exceed those of the higher. We do not know exactly how life first started upon this planet, but we can hardly believe otherwise than that the earliest forms of life were dependent upon their own resources for the synthesis of such special molecular configurations as were required in their metabolism. In the matter of nutrition the earliest forms of life must have been much more like the autotrophic bacteria, making do with simple inorganic compounds, than like any animal alive today. It is natural that we should think of evolution as a progression from the simple to the complex, involving the appearance of new organisms capable of all sorts of activities unknown to their ancestors. But the study of nutrition on the biochemical level presents us with a wholly different picture; the evolution of the higher animals has been marked by a substantial loss of the powers of synthesis which still exist in lower forms of life and which we must believe existed in the remote ancestors of the higher animals. Not only for vital parts of their living machinery, but also for certain raw materials in bulk, the higher animals have become completely dependent upon other organisms.

The conventional carbon and nitrogen cycles of elementary biology suggest that the plants take in simple inorganic substances and use the sun's energy to build them up into complex organic compounds, that the plants are eaten by the animals and that the dead bodies and excreta of the animals are again reduced to simple inorganic substances by the action of bacteria. Undoubtedly this is true in general terms, but a certain qualification is called for; by no means all the plants are eaten by animals. The dead leaves and compost heaps of our gardens bear witness to the fact that even under conditions of intensive cultvation much of the organic matter produced by plants never goes through the bodies of animals at all. If one accepts the classical conceptions of struggle for existence and competition for food substances one cannot but wonder at the immense resources of food material which animals apparently allow to slip through their fingers.

The fact is that while potential food in the form of the bodies of higher plants is almost everywhere abundant, to make effective use of it is not nearly so simple as would appear. If a piece of meat is offered to a sea anemone it will be swallowed, and in the anemone's gut it will be subjected to the action of a proteolytic enzyme secreted by the endoderm cells. The meat will gradually become broken down into a kind of soup of small particles, these small particles will be taken up by the endoderm cells and their digestion will be completed in intracellular vacuoles. Now suppose that instead of a piece of meat we induce the sea anemone to swallow a piece of cabbage. The

5

proteolytic enzyme will penetrate into the substance of the cabbage and will break down the protoplasts of its cells. But the cellulose cell walls will remain and they will prevent all but the soluble products of digestion from escaping. There will be no soup of fine particles set free in the gut cavity to be ingested by the endoderm cells; instead, the fine particles will remain imprisoned in the cellulose cell walls. Given time the action of the proteolytic enzyme might reduce most of these particles to soluble form, and given time the products of digestion might diffuse out of the lump of cabbage, but the whole process would be impossibly slow, too slow even for a sea anemone. To live effectively off the tissues of the higher plants an animal must have at least some means of breaking down plant tissue into small fragments, and preferably also some means of dissolving the cellulose cell walls.

The first of these problems would appear to be a comparatively simple one. The green tissues of plants are tough, but animals are known which can bore holes in much tougher substances such as wood or even rock. Yet in the whole animal kingdom with all its diversity, with all its range of adaptation, this problem of breaking down soft plant tissues has been solved by only three large groups. If one surveys the main groups of the animal kingdom and picks out those which are predominantly and characteristically herbivorous—living on the green tissues (not on the fruits and seeds) of higher plants—one finds oneself left with only three. These are the gasteropod molluscs, certain orders of insects and certain orders of mammals. The gasteropod molluscs make use of that admirable mechanism, the radula, a horny ribbon with teeth, used as a rasp. Insects like the locust have strong sharp mandibles which can both cut and crush plant tissues. The herbivorous mammals have developed molar teeth with corrugated grinding surfaces. The great virtue of this sort of grinding surface is that the corrugations are the result of wear upon three substances of different hardness—enamel, dentine and cement—and therefore there is no tendency for the surface to become smooth with use. These are other devices, of course, such as the stylets of plant-bugs, through which the protoplasmic contents can be sucked out of plant cells, leaving the cell walls intact, but in relation to the general turn-over of material in the carbon and nitrogen cycles these special devices are less important.

The process whereby the food is broken down into fine particles is called trituration. The three main methods of trituration which we have been considering in herbivorous animals greatly facilitate the

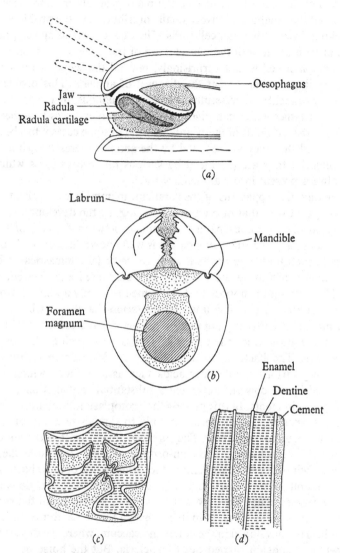

Fig. 1. Means of dealing with plant tissues. (a) Median vertical section through the head of a gasteropod, to show the radular mechanism. (b) Ventral view of the head of a locust, maxillae and labium removed, to show the mandibles. (c) Crown view and (d) vertical section of the molar tooth of a horse, to show the grinding surface.

process of digestion by rendering the plant protoplasm more accessible, but they make a relatively small contribution to the problem of breaking down individual cell walls. The task of extending the process of trituration until the particles are of the same order of size as a single plant cell is, not surprisingly, beyond the scope of most of the grinding mechanisms at the disposal of animals. This problem can only be tackled successfully by chemical methods, by the production of enzymes which can digest cellulose. Such enzymes have been found in the gut fluids of many animals, and in some cases it has been established that they are secreted by the animals themselves; but in the majority they are produced by various micro-organisms which are always present in the alimentary tract.

The digestive apparatus of the ruminant mammals is complicated, as compared with that of man or of the dog, by the development of the relatively enormous paunch or rumen between the end of the oesophagus and the true stomach. As is well known the cow tears off grass in mouthfuls and packs it away more or less unmasticated in the rumen; then at leisure the grass is regurgitated and chewed. It would be wrong to suppose that the rumen is merely a kind of crop or storage reservoir; in fact it is a fermentation chamber in which the cow maintains a thriving population of bacteria and protozoa which live on the grass which the cow eats. This is very much to the cow's advantage. The bacteria are capable of breaking down cellulose, producing various simple fatty acids such as acetic acid which are absorbed by the cow and represent a substantial contribution to its energy requirements. In this process the protoplasmic contents of the cells are set free. The bacteria are able to synthesize many of the vitamins required by the cow. They are also able to synthesize aminoacids from carbohydrate and non-protein nitrogen and build them up into their own protoplasm. Since the ultimate fate of these bacteria is to pass out of the rumen and be digested in the further regions of the alimentary canal, the cow is the residuary legatee of all the synthetic activity which goes on in its rumen. In the horse, fermentation chambers are developed in the colon and caecum, where the digestion of cellulose is again carried out by bacteria. But the horse has not been so clever as the cow, for by having its fermentation chambers at the posterior end of the alimentary canal it is unable to take full advantage of its bacterial flora by digesting their dead bodies.

Among the herbivorous invertebrates fermentation chambers are not so conspicuous a feature of the alimentary canal. They have been described in some insects. But the use of micro-organisms, without the development of any special region to house them, appears to be widespread. Many of the termites, or white ants, live exclusively upon wood. They are able to do this because their intestines contain cultures of flagellate protozoa which take up particles of wood into vacuoles and digest them. The termite does not live directly upon wood but upon the protozoa whose dead bodies it digests.

But the use of micro-organisms in this way is far from universal in insects. Many insects, of which the locust is one, are wasteful feeders, merely crushing the plant tissues with their mandibles so that a few cells are broken open; they eat voraciously and most of what they eat passes out undigested in their faeces.

Effective trituration and the extensive use of micro-organisms are characteristic features of the digestion of herbivorous animals. In carnivores the problems are much simpler since, as we have seen in the case of the sea anemone, the secretion of a strong proteolytic enzyme will ensure an extensive breakdown of animal tissue. Many carnivorous animals are content to swallow the prey whole, but digestion would undoubtedly be more rapid if they performed some mastication upon it. If the prey is large compared with the predator the method of swallowing whole is not entirely satisfactory since the great distension of the body interferes with movement and may place the animal at a disadvantage. This is well known in the case of boa-constrictors and pythons. Some of the invertebrates are able to get over this difficulty by external digestion. The larva of the large water-beetle *Dytiscus* has a pair of long pointed mandibles through each of which runs a tube. The larva plunges the mandibles into its prey and pumps some fluid from its gut into the carcass. The gut fluid contains a strong proteolytic enzyme and after this has been allowed to act the larva sucks back the semi-digested remains of its prey. Star fishes adopt similar methods with mussels, everting their stomachs and applying them to the tissues of the mussel, still partly enclosed in the shell.

Many invertebrates, for example nearly all the lamellibranch molluscs, feed upon small organisms—diatoms, protozoa, bacteria, etc.—suspended in water. The main interest in such microphagous feeders centres upon the filtering devices involved in collecting the food, and with these we shall not be concerned here. We will note only that since the food is already in a finely divided state the problems of digestion are correspondingly simple.

In mammals, and in the vertebrates generally, digestion is wholly extracellular, taking place in the lumen of the alimentary canal. As the food is passed along the canal various enzymes are applied to it in succession and finally the soluble products of digestion—amino-acids from protein, monosaccharides from carbohydrate, and fatty acids and glycerol from fat—are absorbed through the epithelium into the blood or the lymph. One may compare this process with the production line in the manufacture of some article by mass production. The digestive processes of the lower animals are less highly organized. To begin with, digestion is often completed intracellularly, small particles being ingested by the cells of the gut epithelium. This is the case in coelenterates and in nearly all the non-parasitic platy-helminthes. The main disadvantage of intracellular digestion is that it can only go on in the wall of the gut and not at all in the lumen. The animal must either greatly extend the surface of its digestive epithelium—as the platyhelminthes do—or like the coelenterates it must accept the fact that digestion is to be a slow process. In the higher invertebrates such as the insects and the cephalopod molluscs, animals which maintain a high level of activity, digestion is purely extracellular. Intracellular digestion persists among slow-moving animals and in particular among microphagous feeders which are generally slow-moving and in which the already finely divided condition of the food makes practicable its direct ingestion by the cells of the gut.

In the digestive systems of the lower animals, as compared with those of mammals, we do not always find particular stages in the digestive process restricted to particular regions; but we do find a great deal of anatomical differentiation, although it follows rather different lines. The *region of the mouth* is of course distinguished by the special means which the animal uses in swallowing; *Nereis*, for example, has an eversible pharynx, armed with so-called jaws. Salivary glands are by no means uncommon, and, as is likewise the case in mammals, their significance is as much in the mucus they produce for lubrication as in the enzymes they produce for digestion. Following upon the region of the mouth there is very often a non-digestive, purely conducting region, the *oesophagus*. This non-digestive region is sometimes dilated and in this condition is referred to as a *crop*. If by a crop we mean a region of the gut in which food is stored prior to digestion—as in the crop of a bird—the use of the word in the case of many invertebrates is unfortunate. A true crop is characteristic of an animal which feeds intermittently, and of such we could hardly have a better example than the blood-sucking leech. The crop of the leech (Fig. 2) is far larger than all the rest of its alimentary canal, and a single meal of blood takes weeks or months for complete digestion. The function of the crop is purely that of storage, no digestion taking place in it. From the anatomy of the digestive system it is clear that if the blood were to coagulate in the crop it would not be possible for the leech to move it into its minute stomach; but the salivary glands of the leech produce a powerful anti-coagulin which prevents clotting. The crop of the cockroach (or of the snail) is more than a storage organ. Although its walls do not produce any secretion, fluid containing enzymes secreted in more posterior regions of the gut is regurgitated into it. It is therefore a digestive chamber more akin to the stomach of a mammal than to the crop of a bird.

We have seen how very few animals have their mouths provided with adequate means of breaking down their food. We often find that these deficiencies are to some extent made good by the development of organs of internal trituration, or *gizzards*. In the bird the gizzard is a muscular sac with a horny lining and contains small stones or grit with which the food

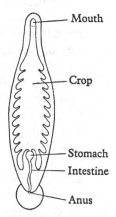

Fig. 2. Diagram of the alimentary canal of a leech, to show the very large crop.

is crushed. In the crayfish the mandibles are not the efficient organs which they are in insects and their action is supplemented in the gastric mill, a region of the foregut provided with chitinous teeth and operated by muscles. The mechanical efficiency of the gastric mill is very low, but its action is assisted by proteolytic enzymes which are passed forwards into it from the mid gut. Associated with these organs of internal trituration we often find filtering devices which allow only small particles to pass beyond. In the cockroach the proventriculus seems to be the homologue of the crustacean gastric mill; but although trituration has been shown to take place in this organ it seems to function mainly as a filter.

The region of the gut from whose walls enzymes are secreted and from which the products of digestion are absorbed is now to be considered. In the annelids it is a tube running straight through the body and the only means used to increase its surface is the development of an inwardly-projecting longitudinal ridge, the typhlosole, so characteristic of transverse sections of the earthworm. In crustacea and in molluscs this region of the gut is complicated by the presence of an organ which is often referred to as the 'liver' or 'hepatopancreas'. But the suggestion of similarity to the liver and pancreas of vertebrates is based upon a misconception of this organ's function and the name *'digestive gland'* is to be preferred. It is generally a paired structure and the ducts by which it opens into the gut branch repeatedly and end blindly in a very large number of small tubes, the alveoli. The alveoli are lined with a typical digestive epithelium and the whole organ is to be looked upon as a means of adding considerably to the area of this epithelium. Semi-digested food in fine suspension is carried into the alveoli. In the molluscs this movement is brought about by cilia, but in the crustacea, which have no cilia, we find that each alveolus has a thin coat of muscle by whose contractions the fluid contents of the digestive gland are kept in motion. Digestion and absorption are virtually completed in the digestive gland by a combination of extracellular and intracellular methods, varying from one type of animal to another. The hepatic caeca of the cockroach probably represent the digestive gland of the crustacea, but no solid food ever penetrates into them. In the cockroach, and in insects generally, the gut secretes a chitinous sleeve, the peritrophic membrane, which retains all the solid matter of the food. This membrane is readily permeable to digestive enzymes and to the soluble products of digestion and its significance appears to be that it protects the digestive epithelium from abrasion by hard particles in the food, a function which in other animals is performed by a copious secretion of mucus.

The *intestine* which extends backwards from the openings of the digestive gland is a very variable structure. In the cockroach the digestive epithelium is continuous from the proventriculus to the openings of the Malpighian tubules and the region defined by these limits is the mid gut, of fairly uniform histological appearance. The intestine which follows the mid gut is a non-digestive, conducting tube. The same is true of the intestine of crustacea. In the gasteropod and lamellibranch molluscs, particularly in herbivores and filter-feeders, the intestine has a glandular epithelium and is long and coiled. No very convincing account has been given of its functions. It has been shown that during their passage through the intestine the faeces are formed into a firm pellet, but animals of other groups seem able to form faecal pellets with intestines of more modest length. One would like to know whether there are any micro-organisms in this part of the gut and, if so, what they are doing.

The *rectum* in most animals is a terminal muscular portion of the alimentary canal whose principal function is to expel the faeces. In terrestrial animals, and in the insects in particular, as we shall see in a later chapter, the rectum is the site of a very important process whereby the faeces are dried before being passed out of the body.

Rather more is known about the properties of the digestive enzymes of the lower animals than about any other aspect of their digestive physiology. If you want to know what enzymes are present in the body of an animal you take it and grind it up with water and sand in a mortar. You then remove the sand and debris by centrifuging and apply various well-established biochemical techniques to the clear fluid. If the purpose of the investigation is thus limited, the invertebrates present no greater difficulties than are met with in the vertebrates. But if you want to know if an enzyme is present in one part of the body and not in another you must of course separate the parts by preliminary dissection; or if you want to know whether an enzyme is present in the fluid contents of a region of the gut or whether it is confined within the cells you have got to collect the fluid and there may not be much of it in a small animal. These difficulties however are not by any means insuperable, and various techniques suitable for rather small quantities have been worked out. Much of the work on insect digestion has been carried out not in test tubes but on waxed glass slides, using small drops of the order of a few cubic millimetres in volume.

In broad outline what we find is that the same sorts of enzymes can be extracted from invertebrates as are known in mammals. The main proteolytic enzymes of the invertebrates are more akin to trypsin than to pepsin. Amylolytic (carbohydrate-splitting) enzymes are generally present but lipolytic (fat-splitting) enzymes are sometimes lacking. On minor matters, such as the optimum pH, we find endless variation, but on the whole the invertebrates appear to be provided with the means of carrying out most of the digestive operations known in vertebrates. There is no doubt that some of the invertebrates can digest substances which are not attacked in the mammalian gut. The ability of the larva of the clothes moth to digest keratin, the protein of wool and hair, is a case in point. Cellulose-splitting enzymes have been reported from many insects and from some other animals but it is difficult to be certain that they are produced by the animal and not by micro-organisms. With a limited number of exceptions, very little has been found in invertebrates that was not already known in mammals.

There is undoubtedly some correlation between the enzyme content of the digestive juices and the nature of the animal's food. Carnivores almost invariably have an extracellular proteolytic enzyme which is much stronger than the corresponding enzyme of herbivores. On the other hand amylolytic enzymes are more active in herbivores. Microphagous feeders, as already mentioned, rely largely upon intracellular digestion. The only extracellular enzyme in the lamellibranch molluscs is an amylolytic enzyme which is remarkable in that it is produced in the form of a transparent gelatinous rod, the crystalline style, Fig. 3. This rod is secreted from a diverticulum of the gut within which it is caused to rotate by cilia, and its projecting end rubs against a horny plate in the roof of the stomach. The string

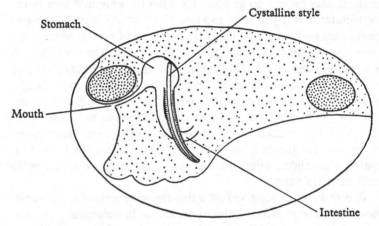

Fig. 3. Diagram of the anterior region of the alimentary canal of a lamellibranch mollusc, to show the crystalline style.

of mucus carrying the filtered particles enters the mouth and becomes wrapped around the style. Probably this device serves to set free the food particles from their mucous entanglement and thus facilitates their ingestion by the cells of the digestive gland.

In the mammal most of the glands which secrete the digestive juices can be activated both by nerves and by hormones. There is a progressive change-over from activation by nerve to activation by hormone as we pass backwards from the mouth. The salivary glands are activated by their nerves only; the initial secretion of gastric juice is brought about via the vagus nerve, but the maintenance of the secretion depends upon hormones; the flow of pancreatic juice in response to stimulation of the vagus is inconsiderable compared with the flow in response to hormones; the flow of intestinal juice occurs in response to local chemical or mechanical stimulation of the epithelium and the nerves are not concerned. By these means the secretion of the digestive juices can be adjusted to the task in hand. Of the coordination of digestion in the invertebrates virtually nothing is known. In the earthworm increased secretion of a proteolytic enzyme follows upon the stimulation of the nerves to the gut. In the snail and in the crayfish it has been shown that the secretory cells of the digestive gland go through cycles of secretory activity; when food is given after a period of fasting the secretory cycles of all the cells become synchronized and their rhythm becomes speeded up. This looks perhaps more like control by hormone than control by nerve. These hints— they are no more—of the existence of some controlling process represent the sum total of our knowledge of this aspect of the comparative physiology of digestion.

2

CIRCULATION

Considering how intensively human anatomy was studied and how great was the fertility of ideas in the ancient world, it seems a little strange that not until the seventeenth century did it occur to anyone that the blood circulated round the body. The Greeks believed that the arteries contained air and that only the veins contained blood. Galen in A.D. 170 showed that the arteries contained blood like the veins and he put forward the ebb-and-flow theory of the blood movements; he supposed that as the heart contracted the blood flowed from it to the extremities of the body and that the blood flowed back to the heart during its relaxation. The circulation of the blood was discovered by William Harvey whose *Treatise on the Motion of the Heart and Blood* was published in 1628. Many people are under the impression that Harvey discovered the circulation of the blood in the sense that he traced the connexion between the arteries and veins and followed the course of the blood upon its circuit. This is not so. Harvey never directly observed the connexion between the arteries and veins and it was not until 1661 that this observation was made for the first time by Malpighi. What Harvey did was to demonstrate in a most elegant manner that the circulation of the blood was a logical necessity. He asked first: is the heart moved by the blood or is the blood moved by the heart? and observing that the heart was hard like a contracting muscle while the blood was leaving it—the phase known as *systole*—and soft like a relaxed muscle while it was filling—the phase known as *diastole*—he decided that the heart was the prime mover in the system. He then made an anatomical study of the heart and its valves and showed that these valves would permit of the flow of blood in one direction only. In the living body he ligatured arteries and veins and saw the blood accumulating upon one side of the ligature and draining away on the other. Finally, he made measurements of the difference in volume between the dilated heart and the contracted heart and calculated that in one hour the heart must pump into the arteries some 500 lb of blood—more than the weight of the whole body. Obviously this would only be possible if the blood came back to the heart, which it could do only through the veins; it therefore followed that there must be a connexion between the arteries and the veins and that the blood must circulate round the body.

The main functions of the blood in the mammal are:
1. Transport of food materials from the gut to the rest of the body.
2. Transport of excretory matter to the kidney from the rest of the body.
3. Transport of respiratory gases between the lungs and the rest of the body.
4. Transport of hormones by which the activities of certain organs are controlled.

These functions are carried out by the mammalian blood system with efficiency and despatch. A portion of blood leaving the left ventricle will pass through the capillaries, say, of the foot, return to the right side of the heart, pass through the capillaries of the lung and be back in the left ventricle in less than half a minute. In view of the distance the blood has to go and the fineness of the vessels through which it has to pass this is not a bad performance. It is made possible by the relatively high pressures developed in the heart and great arteries. In the aorta the average sustained arterial pressure is about 120 mm Hg, rising to about 150 mm Hg during systole and falling to about 80 mm Hg at the end of diastole.

In the course of its circuit round the body the blood passes successively through arteries, arterioles, capillaries, venules and veins before returning to the heart. Each of these different types of blood vessel has special properties which are appropriate to the part that each has to play in the adaptation of the circulation to the needs of the body.

The aorta and the great arteries have thick walls in which there is a great development of elastic fibres. These vessels can stand up to considerable pressures and at the same time their elastic and compliant walls make them capable of accommodating considerable changes in volume without undue change of pressure. If the large arteries had the physical properties of steel tubes the pressure in them would rise so rapidly that the heart would be unable to empty itself completely; conversely, during diastole the arterial pressure would quickly fall as the blood leaked away into the veins. The arteries behave as if they were made of indiarubber rather than of steel and their ability to become distended enables them to function in the same way as the air compression chamber of a reciprocating water pump, which converts the intermittent surges from the pump into a continuous head of pressure.

17

As we pass from the arteries to the arterioles which are about 50μ in diameter we find that the elastic fibres in the wall are progressively replaced by muscle fibres, by whose contraction the diameter of the arteriole can be reduced. These muscle fibres are controlled by the nervous system and by hormones circulating in the blood and it is the state of constriction in the arterioles which determines the rate of blood flow in different organs of the body. From the arterioles the blood, now at the relatively low pressure of 10–30 mm Hg, passes into the capillaries which have thin walls through which exchange of dissolved substances between blood and tissues can take place. Thereafter the blood is gathered up into the venules and then into the veins.

Now we may well ask how it can be that the pressure in the capillaries of the foot is only 10–30 mm Hg if the foot is some 4 ft lower than the heart. A column of blood of this height would exert a pressure of nearly 100 mm Hg. How in these circumstances does the blood find its way back to the heart from the foot? The answer is that although the veins have in themselves no power of contraction they are capable of being pressed flat by the contraction of the leg muscles between which they pass. They are in addition provided with non-return valves at short distances apart. Contraction of the leg muscles squeezes the blood out of the veins and the valves ensure that the blood moves only in the direction of the heart. Thus for the return of blood to the heart the body relies upon the indirect effects of its own muscular activity. The reason why people sometimes faint if they stand to attention for long periods is that in the absence of muscular activity the blood accumulates in the legs and the rate of return to the heart is too slow to maintain the necessary supply to the brain.

The efficiency of the mammalian circulation, the ways in which it can be adjusted to the needs of the different organs and of the body as a whole—and conversely, the degree to which the life of the body is dependent upon the circulation being continuously maintained—are quite without parallel in the lower animals. Indeed many of the lower animals seem able to dispense with a circulatory system entirely. How they are able to do this is an aspect of comparative physiology to which we must now give attention.

The explanation is to be found in the operation of the laws of diffusion. If a substance is dissolved in water and is present in higher concentration in one part of the solution than in another, it will diffuse—by the random movements of molecules—from the region of high concentration towards the region of low concentration. The rate at which it diffuses will be proportional to the concentration gradient, which in simple terms is given by the difference in concentration in the two regions divided by their distance apart. From this it follows that for any given difference in concentration *the rate of diffusion is inversely proportional to the distance over which diffusion takes place.* Now applying this to animals and considering, for example, their oxygen requirements we can readily appreciate that in a small animal an adequate oxygen supply may be maintained by simple diffusion through the body since the distance of any part of the body from the body surface is short, whereas if the same animal were increased in size there would come a time when the slower diffusion of oxygen over the greater distances would no longer be adequate. When this stage is reached the animal can only get bigger if it calls in convection to the aid of diffusion, that is to say, if it develops a circulatory system.

It is not possible accurately to define the size at which a circulatory system becomes essential because so many other factors besides size must be taken into account. The shape of the animal is obviously significant. Weight for weight, if the animal is flattened and leaf-like the distances over which diffusion has to operate between the surface and the interior will be less than if it is more or less spherical. The more active it is the more rapidly oxygen must be supplied, so that for animals of the same general form the inadequacy of diffusion will be felt at a smaller size in an active animal than in a less active one. To take examples. A platyhelminth, such as *Dendrocoelum*, about 1 cm long and about 1 mm thick, seems to get on very well without a circulatory system; the earthworm, larger but certainly not more active, has a circulatory system on a modest scale.

Having thus briefly surveyed the factors entering into the requirement for a circulatory system, let us now consider the ways in which the movement of the blood is brought about.

Nature has a device for moving fluids in tubes which is resorted to by a great many animals including ourselves. This method is known as peristalsis and it consists in causing a wave of contraction, preceded by a wave of relaxation, to travel over the walls of the tube (Fig. 4a). By this method the fluid contents of the intestine are moved along and, although one would not immediately recognize it, this is basically the method which is employed in the vertebrate heart. That this is so can be appreciated by studying the development of the heart's action in the chick. The heart appears at an early stage in development and has the form of a simple tube bent upon itself

(Fig. 4*b*). A wave of contraction starting at the posterior end of the tube sweeps smoothly forwards and this movement is easily recognized as peristalsis. Later in development the region of the future auricle becomes expanded while its walls remain thin, in contrast to the region of the ventricle in which the walls become increasingly muscular. This association of a thin-walled auricle and a thick muscular ventricle to form a two-chambered heart is widespread

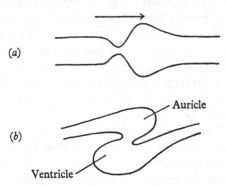

Fig. 4. (*a*) Diagram of a peristaltic wave. (*b*) The heart of a chick at an early stage in development.

among animals of many different groups—it is found, for example, in the heart of the common snail—and we may pause to consider whether there is any particular problem associated with hearts in general to which this arrangement affords a solution.

The more powerful the muscles of the ventricle the faster the blood can be driven through the vessels. But as the ventricle becomes more and more muscular a greater pressure is required to distend it, and for this purpose the low pressure in the veins leading to the heart may be inadequate. It therefore becomes necessary to raise the pressure in two stages rather than in one, and this can be arranged by developing another chamber which is sufficiently thin-walled to be filled by the venous pressure and sufficiently powerful in its contraction to fill the ventricle.

Another feature of the heart in many groups of animals is that it is enclosed in a cavity called the pericardium, usually a part of the general body cavity set aside, as it were, for the heart's private use. Since the heart undergoes regular changes in volume it is obvious that it must not be crowded and jostled by the other viscera. The provision of space in which the heart is free to work would appear to be the primary significance of a pericardium. But in some animals it may have a further significance in relation to the filling of the heart. The pericardium is enclosed by a tough pericardial membrane and this is continuous with the connective tissue sheets which cover other viscera and attach them together. The volume of the pericardium cannot be diminished without pulling upon other structures and therefore anything which tends to diminish the volume of the peri-cardium will tend to set up a slight negative pressure inside it. Now the contraction of the ventricle and the expulsion of its contents is to all intents and purposes a reduction of the volume of the pericardium and if a negative pressure is set up, this will be transmitted through the thin wall of the auricle and will cause the blood to flow in from the great veins outside the pericardium. In this way successful filling of the auricle can be combined with a low venous pressure.

In the chick we can follow the stages by which the simple tube undergoing peristalsis becomes replaced by the two-chambered heart, and we see how as this development proceeds the auricle and ventricle come to contract as units in succession, the smooth passage of the peristaltic wave being replaced by two separate contractile efforts. In the mammalian ventricle there are special arrangements which ensure the contraction is initiated at exactly the same instant in all parts of the muscle. If it were otherwise the pressure set up by the contrac-tion of one part of the ventricular muscle would stretch and rupture the part that was still relaxed.

Having regard to these problems of hydraulics it should not then surprise us if in circulatory systems of a primitive type working at low pressure we see peristaltic waves passing along tubes, whereas in more advanced systems working at higher pressures we should expect to find chambers in which contraction takes place simultaneously at all points in the muscle.

The first thing that strikes us when we survey those animals which do not possess blood vessels is that they are by no means all of that small size which considerations of diffusion would lead us to expect. Among the coelenterates the sea anemones are some inches across and jellyfishes are known to grow to a diameter of 6 ft. The existence of such animals appears at first sight to make nonsense of our theorizing. But there is something more to be said on this subject. The sea anemone may be some inches across but this does not mean that the thickness of its tissues is to be measured in inches. If we make sections of a sea anemone we find that it has the structure of a relatively thin-walled bag fitted with many radial partitions, called mesenteries. Sea water enters freely through the mouth and bathes the surfaces of the mesenteries. Apparent thickness and solidity in the sea anemone are achieved by this device of folding the tissues into thin flat plates, which at the same time ensures that the living tissue is nowhere very thick and that diffusion paths remain short. If we examine the jellyfish we see that more than 90 per cent of its bulk is due to the non-living jelly or mesogloea and that the living tissues

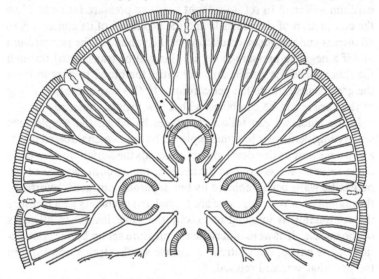

Fig. 5. Jellyfish, to show the gastro-vascular system.

are confined to a thin layer covering the external surface and lining the gut. And what is more, the jellyfish *has* got a circulatory system in its gut. A system of fine canals runs from the central stomach to the circular canal near the margin of the bell (see Fig. 5). These canals are provided with cilia by whose action a perfectly definite circulation is maintained. Sea water enters by the mouth, passes outwards by the eight unbranched canals, returns by the eight branched canals and leaves by the grooves in the oral arms. A circulatory system of this type is sometimes called a gastro-vascular system.

Another group of relatively large animals lacking a proper circulatory system is seen in the echinoderms. One of the strange things— and there are many strange things—about the echinoderms is that they have a perfectly good set of vessels, but do not circulate the fluids in them. In the starfish the arrangement of the vessels takes the form of a ring round the mouth with an extension down each arm.

There are altogether three different systems of tubes in the starfish, all having this same general plan. One system, the so-called water-vascular system, is freely open to the exterior by a sieve-plate and sea water is driven inwards by cilia in the pores. But there is no evidence that the water-vascular system—or any other system—in echinoderms performs any of the functions which we associate with the circulatory systems of other animals and it is reasonable to regard the echinoderms as having no circulatory system. How can they, at their relatively large size, get along without one? First, the amount of living tissue in an echinoderm is by no means commensurate with the animal's bulk; a sea urchin is something like a hollow ball of calcium carbonate filled with sea water. Second, the echinoderms are extraordinarily sluggish animals, with a long history of sessile ancestors. As sessile animals their way of life would not demand a high metabolic rate; as free-living animals not yet having evolved a proper circulatory system they must of necessity go about their business in slow time.

Looking for animals in which the beginnings of a circulatory system are recognizable we turn next to the annelids. The earthworm, although it may certainly be classed among the lower animals, exhibits certain anatomical complications which make it less suitable to take as our next example than the marine worm *Nereis*. In *Nereis*, as in the earthworm, there is a dorsal vessel and a ventral vessel running the whole length of the body, and these two main vessels are connected in each segment by a system of transverse loops. The arrangement of these loops is shown in Fig. 6.

In the dorsal vessel the blood is driven forwards by peristalsis. This is perhaps not at first obvious because the wave length of the peristaltic wave, instead of being short in relation to the diameter of the vessel as represented in Fig. 4a, is much longer, being comparable with the length of the whole worm. But close observation will attest that the process is of the same nature, that is, a wave of contraction preceded by a wave of relaxation. In the transverse vessels, the intestinal vessel and the ventral parapodial vessel, peristaltic waves are also seen. They start at the central end and pass outwards towards the parapodia, dying out after a short distance.

As far as has been established the course of the blood is regular and consistent in the following parts of the system. In the dorsal vessel it is driven forwards and is also driven downwards into the intestinal network. In the ventral vessel, which is non-contractile, it flows passively backwards. From the intestinal network it is driven outwards by the intestinal vessels and from the ventral vessel it is driven outwards by the ventral parapodial vessels towards the parapodium. It returns from the parapodium mainly through the dorsal parapodial vessels which open into the dorsal vessel. But in the vessels of the parapodium, which form a richly anastomosing network, the blood does not move consistently in one direction. The blood vessels of the parapodium are actively contractile but their activity takes the form

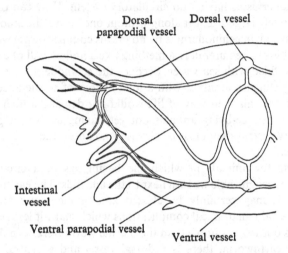

Fig. 6. The blood vessels of *Nereis*, as seen in transverse section.

of a general dilatation—a sort of blushing—followed by a contraction which drives the blood out again. The movement of blood in the parapodium is thus largely of the nature of ebb-and-flow. This, it will be remembered, was the picture Galen had of the mammalian circulation and blood movements of the ebb-and-flow type are called galenic.

One must be struck by the vagueness and apparent lack of coordination in this system. For example, the peristaltic waves in the intestinal vessel and the ventral parapodial vessel are not synchronized and they may be more powerful in one vessel than in the other, so that in the short vessel which connects them the blood may flow ventrally at one time and dorsally at another. Similarly there is no regularity about the direction of flow between the ventral vessel and the intestinal network.

Now it is clear from all this that we should be led astray in trying to understand the circulation in *Nereis* if we approached it with any fixed ideas, based upon the higher animals, of what a circulation should be like. Here is no localized pump driving blood through a system of passive tubes; instead, we find that the power of contraction is widely extended throughout the vascular system. Here is no round-the-body circulation, from heart through arteries, capillaries, veins, and back again; instead we can discern a definite circulation of blood round the body in the sagittal plane, i.e. forwards in the dorsal vessel and backwards in the ventral vessel, and in each segment a more or less independent circulation in the transverse plane which is much obscured by ebb-and-flow movements in the parapodia.

The vascular system of the cockroach shows a superficial resemblance to the vascular system of *Nereis*, for in both there is a dorsal contractile vessel which drives the blood forwards, But as we shall see in a moment in the cockroach—and in arthropods generally—the relation of the vascular system to the rest of the body is radically different from the relation we find in worms and in vertebrates, and these differences in morphology are associated with certain differences in hydraulics.

At a fairly early stage in his study of zoology the student may wonder why it is that when he opens the body of an earthworm he finds the internal organs lying in a cavity which is called the coelom, whereas in the cockroach the corresponding cavity is called the haemocoel. To be told that the haemocoel contains blood whereas the coelom does not is no answer since the blood of the insect has the same appearance as the coelomic fluid of the earthworm. It is only when we study and compare the development of annelids and arthropods that we can realize the true nature of the differences in their body cavities.

Rather than get involved in the details of development in particular animals, let us generalize, at some slight expense in accuracy, by saying that the cleavage of the fertilized egg gives rise to a hollow sphere of cells, known as a blastula, whose cavity is known as the blastocoel or primary body cavity. In the process of gastrulation one end of the blastula becomes pushed in and in due course this in-pushed portion will become the gut. We can now recognize by their position the ectoderm of the outer surface and the endoderm which lines the gut. Masses of cells, the mesoderm, are now proliferated into the blastocoel and in the interior of these masses the cells come apart to form another cavity. This is the secondary body cavity or coelom, and being entirely surrounded by mesoderm it is completely isolated from the blastocoel. The relation between the primary and secondary body cavities is illustrated in Fig. 7a.

Up to this stage what has been described would apply equally to an annelid and to an arthropod, but from now on we find that development takes different courses in the two groups. In the annelids the

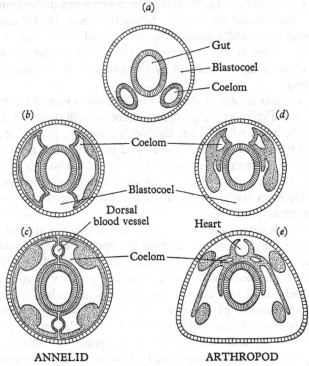

Fig. 7. Diagrams to show the relations of the cavities of the body and their condition in annelids as compared with arthropods.

coelom begins to increase in size, eventually to become the body cavity of the adult, and it squeezes the blastocoel almost but not quite out of existence. As shown in Fig. 7b, c there are two regions where the blastocoel persists and these become the dorsal and ventral blood vessels. This study shows us that the cavity of the vascular system is morphologically identical with the primary body cavity or blastocoel; the word haemocoel is applied to the blastocoel when it has attained the condition of being a cavity through which fluid is circulated. We may therefore write:

haemocoel = blastocoel + circulating blood.

But in the arthropods the coelom does not enlarge at the expense of the blastocoel. It remains small and persists mainly as the cavities of the genital organs. The mesoderm increases in thickness as it does in the annelid, and shows the same upward growth; and the arthropod heart is formed out of two halves in much the same way as the dorsal vessel of the annelid. But there is this fundamental difference that whereas in the annelid the wall of the dorsal vessel separates haemocoel from coelom, in the arthropod the wall of the heart merely separates one portion of the haemocoel from another. The heart and all the rest of the viscera lie free in the haemocoel and are bathed in blood instead of lying free in the coelom and receiving a piped supply of blood. In the annelids and vertebrates the body cavity is formed by the coelom and the blood system is of the closed type, whereas in the arthropods and in the molluscs the body cavity is formed by the haemocoel and the blood system is of the open type.

The adult arthropod condition as seen in transverse section is shown diagrammatically in Fig. 8. In a primitive arthropod the heart is a long muscular tube. In each segment of the body it has a pair of openings, the ostia, guarded by valves, and at its anterior end it is prolonged into a limited arterial system. Blood enters through the ostia and the contraction of the heart takes the form of a peristaltic wave which sweeps forward over it. The blood is distributed to the head, sometimes also to other parts of the body, by the arteries and then escapes into the general haemocoelic body cavity through which it makes its way back to the heart. Such direction as is imparted to the flow of blood through the body cavity is due not so much to the provision of special channels as to the presence of membranes and baffles round which the blood must make detours.

In the crayfish the heart is a compact organ lying at the posterior end of the thorax, with well-developed arteries running forwards and backwards. Seeing that the insects are in every other way the most advanced arthropods it may seem strange at first that the insect heart is on the same level as the heart of a primitive crustacean. The explanation is of course that in the insect, owing to the development of the tracheal system, the blood has no respiratory function; for the other functions of the blood a relatively poor circulation is adequate.

In mammals, as we have seen, the blood returns to the heart at a pressure which although low is yet sufficient to fill the first chamber or auricle. But since the arthropod heart is freely suspended in the blood it is impossible for the heart to be distended by internal pressure. How then is it filled? From Fig. 8 it can be seen that the heart lies in

27

a pericardial space which is partially separated from the general haemocoelic body cavity by a horizontal sheet of tissue, the pericardial membrane. The heart is held in position by elastic ligaments which suspend it from the dorsal body wall and attach it to the pericardial membrane. The contraction of the heart stretches these elastic ligaments and during relaxation the diverging ligaments pull the walls of the heart apart and cause the blood to flow in. In some arthropods, of which the cockroach is one, the action of the elastic

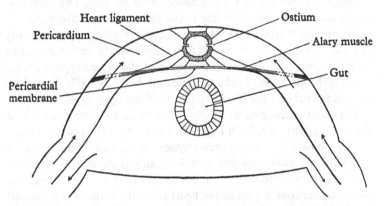

Fig. 8. Diagrammatic transverse section of an insect, to show the relation of the heart to the body wall and body cavity.

ligaments is reinforced by muscles, known as the alary muscles, which run in the pericardial membrane from the dorso-lateral body wall to the ligaments of the heart. By their contraction, which occurs during the relaxation of the heart itself, the tension on the ligaments is increased.

It will be appreciated that this method of distending the heart by means of elastic ligaments is only available to an animal in which the heart is partly surrounded by a rigid skeleton. The snail, as mentioned, has an open blood system and although it has a rigid external skeleton in the form of its coiled shell this does not serve for the attachment of elastic ligaments as in the arthropod. The snail's heart, however, is not suspended in the blood like the heart of the arthropod. Like most hearts it is enclosed in a pericardium and in this case the pericardium is a coelomic cavity as in the vertebrates. It is therefore open to the snail to take advantage of any negative pressure which may be set up in the pericardium by the expulsion of blood from the ventricle. Whether or not this is the true explanation of the filling of the snail's heart must remain undecided in the absence of experimental evidence. But the easily observed fact that the filling of the auricle is synchronized with the emptying of the ventricle suggests that it may be.

The sustained arterial pressure, characteristic of mammals, depends not only upon the powerful contraction of the ventricle but also upon the elasticity of the main arteries and the peripheral resistance of the finer vessels. In insects like the cockroach the heart is continued anteriorly as a simple aorta which opens widely in the region of the head. There is therefore practically no peripheral resistance, no sustained arterial pressure, and the flow of blood is intermittent. This is easy to see in small transparent insects like mosquito larvae. The blood contains haemocytes in suspension and its movements can be followed by watching the haemocytes under the microscope; with

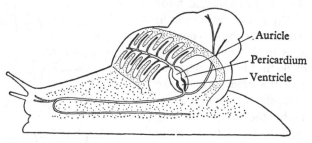

Fig. 9. Optical section of a snail, to show the circulatory system.

each contraction of the heart the haemocytes in the general haemocoel make an abrupt backward movement and then stay motionless.

In the annelids the peripheral blood vessels are quite fine but, since they are themselves contractile and largely responsible for the movements of the blood they contain, peripheral resistance has very little meaning. The thin walls of the larger vessels and the obviously peristaltic nature of their contraction are not suggestive of a high blood-pressure. Some measurements have been made of the pressure in the dorsal vessel of *Nereis* and as one would expect it is very low, about 1·0 to 1·5 mm Hg above the general coelomic pressure when the worm is at rest, rising to about 6·0 mm Hg when the worm is active.

Among the invertebrates the highest pressures have been recorded in the octopus and its allies. The spacious sinuses characteristic of the open blood system in other molluscs are here restricted to definite channels; in fact we are able to recognize arteries, arterioles, capillaries, venules, and veins as in mammals and there is some evidence that the flow of blood through the peripheral vessels is under nervous control.

Against the resistance offered by the peripheral vessels the heart is able to set up and sustain a very reasonable pressure, which varies from about 45 mm Hg to about 30 mm Hg according to the phase of the heartbeat. Another interesting feature of the circulation in these animals is the presence of branchial hearts in addition to the normal systemic heart. In fishes the blood leaving the heart has first to pass through the capillaries of the gills before being distributed to the organs of the body. As a result the pressure in the dorsal aorta is some 30 per cent lower than in the ventral aorta. In the mammals and birds the difficulty of two resistances in series is met by having two pumps; the heart is divided into two halves, one driving the blood through the lungs and the other driving the blood through the rest of the body. In molluscs the blood returning from the body to the heart has to pass through the vessels of the gills (called ctenidia in molluscs) and the same problem is thus presented in a different form. In the octopus and its allies it is solved by the development of two branchial hearts, one at the base of each ctenidium (Fig. 10).

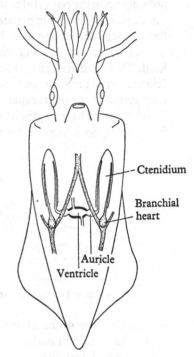

Fig. 10. Respiratory organs and circulatory system of *Loligo*. The pericardium is not shown.

It is characteristic of the heart, in whatever group of animals we study it, that it continues to beat after all nervous connexions with the rest of the body have been severed. The beat of the heart is initiated within the heart itself. In vertebrates and molluscs the heartbeat continues after all nervous tissue within the heart has been extirpated or paralysed with drugs; when the heart-beat is initiated in the heart muscle we speak of the heart as myogenic. In arthropods the heart-beat is initiated and conducted to the heart muscle by nerve cells situated in the heart wall; such hearts are said to be neurogenic.

In the mammalian heart the beat is initiated at the so-called sinuauricular node, where the great veins enter the right auricle. This pace-maker receives a double innervation, from the vagus nerve and from the sympathetic nervous system. If the vagus nerve is stimulated the heart-beat is slowed down and conversely it is accelerated by stimulation of the sympathetic. Thus while the beat of the heart is initiated locally the rate of beat is under the control of the central nervous system through the balance which is struck between the activities of the vagus and of the sympathetic. In the wall of the right auricle there are sensory nerve endings which are stimulated by stretching of the auricular muscle. From them nerve fibres run up in the vagus nerve trunk to the brain and activity in these sensory nerves causes acceleration of the heart rate. Thus if there is an increase in the rate of return of blood to the heart—such as would be a result of muscular activity—the increased rate of filling of the auricle tends to speed up the heart. Another set of sense organs is present at the base of the aorta and these are stimulated by increase in blood-pressure; when the pressure in the aorta rises above its normal level the heart is slowed down. Through these mechanisms the activity of the heart is adjusted to the needs of the body.

In most of the invertebrates there are nerves running from the central nervous system to the heart and stimulation of these nerves causes alteration of the heart rate. Both acceleratory and inhibitory nerves have been found in molluscs and in arthropods. The means of adjusting the heart-beat in relation to the needs of the body appear to exist; but at present we have no idea of the fineness of such adjustment nor of the sensory channels through which it may be brought about.

In approaching the vascular systems of the lower animals and in seeking to compare them with our own we should always bear in mind that in so far as the lower animals are smaller and less active that we are, so are they less dependent upon efficient circulation of their body fluids. At first sight their arrangements will seem to us regrettably haphazard. We are accustomed to think of a vascular system as a circulatory system and of a circulatory system, either living or non-living, as a system of pipes provided with a pump, having a regular and continuous flow in one direction only. The circulation of some invertebrates is more in the nature of a process of stirring. In the snail, for example, the movements of the blood which are brought about when the animal retires into its shell are very much more extensive and violent than anything which the snail's heart can produce. Even when the blood is confined to a system of pipes its flow is not always one way. Very often it flows into blindly ending vessels which then contract and drive it out again, or where several alternative paths are open to it it may flow in one direction at one time and in the opposite direction at another.

In terms of evolutionary theory we may reasonably suppose that there is no advantage (upon which natural selection might act) in possessing a circulatory system which can provide supplies faster than they can be used. A great deal of energy is expended by a mammal in sending its blood rushing at speed around the body and forcing it through fine vessels, but where a high level of sustained activity requires that this should be done, natural selection also requires that it should be done efficiently. Where a relatively sluggish circulation is adequate very little energy is wasted in bringing it about and the selection pressure in favour of efficiency is correspondingly low.

Apart from that immediate advantage of shortened diffusion paths which the possession of a circulatory system confers upon an animal we must not lose sight of a very much more fundamental consequence of the appearance of the circulatory system in evolution. A circulatory system makes possible the localization of physiological functions and the devlopment of the special organs which carry them out. A special respiratory organ is of little use to an animal which lacks the means of bringing it into physiological relationship with the rest of the body. To illustrate the consequences to the body organization of the absence of a circulatory system we shall return for a moment to the platyhelminth. We have seen how, at this small size, simple diffusion is adequate for the exchange of oxygen and carbon dioxide between the tissues and the outside world. The rate of diffusion depends not only upon the concentration gradient but also upon the size of the molecule, small molecules diffusing more rapidly than large ones. Oxygen and carbon dioxide are relatively small molecules; but even after the processes of digestion have been at work the molecules of food substances are by this standard large. We see this reflected in the structure of the platyhelminth's gut. Instead of being a simple tube or bag it is branched and its branches extend throughout the body. In the platyhelminth it appears that if food substances cannot reach the more distant tissues by diffusion the gut itself must go there. Similarly the excretory system (so-called) and the reproductive system are spread throughout the body. The whole principle of division of labour and the development of organs with special function, upon which the physiological efficiency of the higher animals appears to depend, must wait upon the appearance of the circulatory system which, complementary to the nervous system, affords a means of integrating their separate activities.

3

RESPIRATION

In Chapter 1 we noted that the energy requirements of the animal body were met by the oxidation of relatively complex organic compounds and involved the uptake of oxygen from the atmosphere and the elimination of carbon dioxide. It is these oxidative processes at the cellular level, common to all animals, that are the fundamental phenomena of respiration; the respiratory movements, to which the original meaning of the word was restricted, are ancillary processes and not of universal occurrence.

Since this book is about physiology and not about biochemistry we shall not go into details of the sequences of chemical transformation undergone by the food substances during their oxidation. But rather than leave this part of the canvas completely blank we shall sketch in the lines which these transformations follow and indicate the principles to which they conform.

It will help if we begin by distinguishing between two principal forms of energy: heat (or, more correctly, thermal energy) and what we shall here call 'useful energy'. Heat is not a form of energy which is useful to organisms; they cannot use it for the performance of work or for synthesis. Useful energy, which provides for the work done by our muscles, is supplied to the contractile mechanism in the form of an organic compound widely known by its initials, ATP, which stand for adenosine triphosphate; the splitting of this molecule into a molecule of adenosine diphosphate (ADP) and a molecule of phosphoric acid makes useful energy available on the scale of 7,000 calories per mole of ATP. It is a remarkable fact that throughout the whole range of organisms, from bacteria to mammals, ATP is a universally acceptable source of useful energy. The whole complex series of processes which we call respiration may be said to culminate in the production of ATP.

Another widely acceptable source of useful energy, acceptable because its useful energy is readily made available in the form of ATP, is glucose. The oxidation of one mole of glucose to carbon dioxide and water is able to provide 685,600 calories of useful energy. One mole of glucose should therefore be capable of providing something approaching 100 moles of ATP. In fact it provides only 38 moles

33

of ATP, equivalent to 266,000 calories of useful energy, the rest of its useful energy being converted into heat and wasted. Why does this happen?

A single reaction in which one molecule of glucose, six molecules of oxygen and 100 molecules each of ADP and phosphoric acid all took part simultaneously would be unthinkable to a chemist. A chemist is used to dealing with bimolecular and trimolecular reactions, but poly-molecular reactions of this order are widely improbable because of the unlikelihood that so many reactants would all be present together exactly in the right places at any given moment. What happens in living organisms is that the glucose molecule is broken down by stages in a series of simple reactions, each of which is mediated by an enzyme. At each step some of the useful energy becomes available. In some steps there is enough available for the synthesis of a molecule of ATP; in others the useful energy liberated is less than 7,000 calories per mole and it simply goes to waste as heat. In the process of converting one mole of glucose (6 carbon atoms) into two moles of pyruvic acid (3 carbon atoms each), a process which involves some ten separate reactions, something like 37,000 calories of useful energy are liberated and only two moles of ATP are synthesized.

The food materials ingested by animals comprise proteins, carbohydrates and fats. The large molecules of these substances are broken down with the help of enzymes into small molecules containing 2 to 5 carbon atoms. These processes of breakdown, though they liberate some useful energy, do not provide any ATP, except in the case of glucose as described in the last paragraph. At this stage the metabolic pathways converge, both in the chemical sense and in the sense that the remaining reactions all take place within certain intracellular organelles, the mitochondria.

The next steps, known as the Krebs cycle, involve the removal of carbon atoms as carbon dioxide. This does not require the participation of molecular oxygen; instead, the oxygen is taken from water and the hydrogen atoms are temporarily accommodated by special molecules which serve as hydrogen carriers. These steps are accomplished with little change in energy, and again only two molecules of ATP are gained per molecule of glucose. But since the start of the whole series of reactions no less than 12 pairs of hydrogen atoms have accumulated in charge of the carriers. These still await oxidation and it is clearly in this final process that the main yield of ATP must be realized. It is scarcely possible to describe in a few sentences what happens to these hydrogen atoms before they are finally united with oxygen to form water. In principle, each hydrogen atom is temporarily dissociated into a hydrogen ion and an electron. The electrons are passed along a chain of electron carriers, of which the various cytochrome pigments are the best known, giving up useful energy at each step. From the process of electron transport a total of 34 molecules of ATP is harvested for every molecule of glucose consumed.

From the purely biochemical point of view interest in respiration begins when oxygen dissolved in the body fluids enters into chemical reaction with the enzyme systems inside the living cells and ceases when the carbon dioxide and water have been produced. Where these cells lie in the body or in what sort of animal they lie is not of immediate importance. The physiologist, however has to consider not only how the oxygen is used in the cells but also how it gets there and how the carbon dioxide gets away, and this involves the study of two further processes, first, the exchange of gases with the external medium at the respiratory surface, and second, the transport of gases through the body between the respiratory surface and the respiring tissues. For the physiologist it matters a great deal what sort of animal he has to do with and where it lives.

We will begin by considering the exchange of gases in the human lung. During the early part of the present century opinion was divided about the mechanism of oxygen uptake. One school maintained that the passage of oxygen into the blood could be accounted for by simple diffusion; the other school believed that diffusion alone was inadequate and that the lung epithelium actively secreted oxygen into the blood. This controversy is now settled and all are agreed that it is unnecessary to postulate any active secretory process, either for oxygen or for carbon dioxide. The same goes for all animals, vertebrate and invertebrate, in which respiratory exchange has been studied.

In order to show that exchange of oxygen in the lung can be accounted for by diffusion alone we have first to show that the concentration of oxygen in the air is greater than its concentration in the blood (and conversely for carbon dioxide). Here we meet the difficulty that in the cavity of the lung oxygen is a component of a gas mixture while in the blood it is combined with haemoglobin. If we are to compare the concentrations we must first express them in the same units. The concentration of any gas in a mixture of gases is given by its partial pressure; thus the partial pressure of oxygen in the atmosphere is approximately 21 per cent of 760 = 160 mm Hg.

We can express the concentration of oxygen in the blood as *the partial pressure of oxygen in a gas mixture which is in equilibrium with the blood.* We call this the oxygen tension of the blood and it is, of course, expressed in mm Hg. There are various ways of determining the oxygen tension of blood. One is to take a small bubble of air, expose it to a large volume of blood until equilibrium is established and then analyse the bubble. In man the oxygen tension of arterial blood is 95 mm Hg and that of venous blood is 40 mm Hg. In the air drawn from the depths of the lungs—the alveolar air—the partial pressure of oxygen is 105 mm Hg. It therefore follows that oxygen will tend to diffuse from the alveolar air into the blood. If we had found that the oxygen tension of the arterial blood was greater than 105 mm Hg we would have had to conclude that oxygen was actively secreted into the blood by the lung epithelium.

For the exchange of gases between the animal and the external medium it naturally makes a great deal of difference whether the medium is air or water. Air contains 210 c.c. of oxygen per litre. At ordinary temperatures fresh water, when in equilibrium with air, contains about 7 c.c. of oxygen per litre and sea water only about 5. Thus a litre of air will provide oxygen to support life for very much longer than a litre of water. On the other hand the solubility of carbon dioxide in water is very much greater than the solubility of oxygen, and what is more, most natural waters contain a small amount of carbonate with which carbon dioxide will combine to form bicarbonate. From the point of view of getting rid of carbon dioxide water is a better external medium than air. If a frog is kept immersed in water up to the head it takes in most of its oxygen direct from the air via its lungs and gets rid of most of its carbon dioxide to the water via its skin.

Another very important difference between water and air is that in water the rate of diffusion of oxygen is many thousand times less than its rate of diffusion in air. This fact, taken in conjunction with the low solubility of oxygen in water, has important consequences. In the previous chapter we saw how the slowness of diffusion through the tissues of the body made it necessary for animals to supplement diffusion with convection. For precisely similar reasons water-living animals have to supplement diffusion with convection outside their bodies. If still water is allowed to remain in contact with the respiratory surface the oxygen in the nearer layers is quickly used up and the rate at which it is renewed by diffusion from more distant layers soon becomes inadequate to meet the animal's requirements. An animal which breathes water must maintain a current over its respiratory surface whereas an animal which breathes air can often get away with diffusion alone. The earthworm, for example, is not known to take any special action to renew the air in its burrows, but *Nereis*, when it is in its tube, maintains a current of water through the tube by undulatory movements of its body. Crayfishes and crabs have a gill chamber under the carapace and have a special paddle-shaped limb, the scaphognathite, which causes water to enter the gill chamber by small openings around the bases of the legs and drives it out forwards to the side of the head. The common shore crab *Carcinus* can live equally well in water or in air and provides us with an excellent illustration of the point under discussion. If its scaphognathites are immobilized by cauterization the oxygen consumption of the crab in sea water falls and it soon dies, but in air it maintains its oxygen consumption and suffers no apparent inconvenience from the operation. Most insects, as we shall see later, rely solely upon diffusion for the renewal of oxygen in their tracheae.

This discussion of the relative merits of air and water as respiratory media shows air to have a substantial superiority by virtue of greater oxygen content and greater rate of diffusion. But we have not finished yet; the design of the respiratory organ has still to be considered. The gills of a fish consist of a set of gill bars from each of which a large number of long thin-walled gill filaments trail in the respiratory current, presenting a large respiratory surface. But if the fish is lifted out of the water, surface tension causes the gill filaments to clump together and the effective area of the respiratory surface is drastically reduced. It is no easy problem to produce an array of filaments or a branched tree-like structure which is at the same time sufficiently strong to maintain its shape against surface tension and against gravity and sufficiently thin-walled to allow easy passage to the gases. Most animals which breathe air avoid the difficulty by having the respiratory organ in the form of an invagination of the surface rather than an outgrowth from it. The respiratory surface is increased by further invagination of the walls on a smaller scale so that the organ comes to contain a very large number of small pockets, or alveoli, leading to a common chamber or duct with a restricted opening to the exterior. Besides having the advantage of greater structural strength in the mutual support afforded by the walls of the alveoli, the invaginated respiratory organ suffers less water loss by evaporation since the air inside it is moist—and to terrestrial animals water loss is important. But, unfortunately, if water vapour accumulates in the lung so also does carbon dioxide. As we shall see in a moment high concentrations of carbon dioxide have a disastrous effect upon the transport of oxygen by the blood. The partial pressure of carbon dioxide in the atmosphere is about 0·2 mm Hg; its partial pressure in the alveolar air of man is 40 mm Hg; in natural waters one does not find a carbon dioxide tension of 40 mm Hg except in the very foulest swamps.

In the outcome, then, it appears that there is nothing to beat good clean water—if you can get it. But there lies the trouble. The sea is reliable on the whole, but fresh waters very definitely are not. The oxygen content of fresh water is maintained by aeration at the surface and by the photosynthetic activities of plants. During drought rivers may become stagnant pools and in many lakes the bottom waters are devoid of oxygen at certain seasons of the year. The relative merits of air-breathing versus water-breathing are to be assessed on a broad ecological basis and not merely on grounds of physiological efficiency. The supply of oxygen for water-breathing animals is precarious, the supply of oxygen for air-breathing animals is assured. The terrestrial animal can count upon its 21 per cent of oxygen being available the world over, irrespective of the vagaries of weather. It is no easy matter to change over one's respiratory mechanism from water-breathing to air-breathing, but to do so is to invest in a gilt-edged security.

We turn now to the transport of gases through the body. Small animals without a circulatory system rely upon diffusion through their tissues. In many animals of somewhat greater size the gases are carried about in simple solution in the blood. Now as we saw a moment ago carbon dioxide is very much more soluble in water than oxygen and can also be carried in chemical combination as bicarbonate. The carbon dioxide tension in our own tissues is about 50 mm Hg; at a partial pressure of 50 mm Hg the blood plasma (without the corpuscles) will carry about 55 c.c. of carbon dioxide per 100 c.c. plasma. But with oxygen it is altogether different. In equilibrium with the alveolar air, at a partial pressure of 105 mm Hg, the blood plasma will carry less than 0·4 c.c. oxygen per 100 c.c. plasma. The carbon dioxide-carrying capacity of the blood would be of quite a different order from the oxygen-carrying capacity were it not for the presence of haemoglobin, by which the oxygen-carrying capacity is raised from 0·4 vols. per cent to 20 vols. per cent.

Haemoglobin is the most important member of a class of substances known as blood pigments, which are capable of entering into a loose reversible combination with oxygen. It is found in all phyla above platyhelminthes—though of course not in all species. It is bright red when oxygenated and purple-red when de-oxygenated. The only other blood pigment which we will take note of is haemocyanin which is found only in molluscs and crustacea; it is blue when oxygenated and colourless when de-oxygenated. The significance of a blood pigment is that it increases the oxygen-carrying capacity of the blood. A few examples are given in the following table:

Animal	Blood pigment	O_2 capacity, vols. per cent
Planorbis (water snail)	haemoglobin	0·9–1·5
Nereis	haemoglobin	11·5
Man	haemoglobin	20
Helix (common snail)	haemocyanin	1·1–2·2
Astacus (crayfish)	haemocyanin	2·4
Loligo (cephalopod)	haemocyanin	3·8–4·5
Sea water for comparison		0·5

When the blood pigment is exposed to a high oxygen tension in the respiratory organ it takes up oxygen and when it is exposed to a low oxygen tension in the tissues it gives the oxygen up again. The relation between oxygen tension and the amount of oxygen combined with the pigment is not a straight line but an S-shaped curve, and this is a matter of some physiological importance.

For various reasons which need not be entered into now we shall find it easier to study the properties of a blood pigment in the squid Loligo than in ourselves. Fig. 11 shows the relation between oxygen tension and the amount of oxygen carried in the blood. We call this the oxygen dissociation curve[1] of the blood. The dissociation curve indicates that at an oxygen tension of 140 mm Hg the blood is fully saturated with oxygen and contains 4 vols. per cent. The straight line drawn from the saturation point to the origin shows what would happen if the relation between oxygen tension and oxygen content was one of simple proportion. Now consider the animal living in sea water of reduced oxygen tension so that the oxygen tension of its arterial blood is only 70 mm Hg. At this tension the blood will carry 2·95 vols. per cent; if the relation was one of simple proportion it would carry only 1·95 vols. per cent. Let us further suppose that the oxygen tension of the venous blood is 30 mm Hg, at which tension the blood will carry 0·65 vols. per cent as compared with 0·8 vols.

[1] It may be as well to point out that in the conventional dissociation curves of physiological text-books the percentage saturation of the blood, not the oxygen content in vols. per cent, is plotted against the oxygen tension.

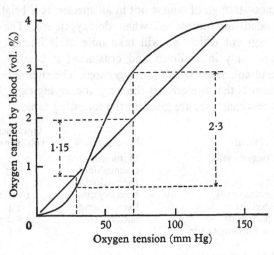

Fig. 11. Oxygen dissociation curve of the blood of *Loligo*, to show how the S-shape implies a greater oxygen capacity over the middle range.

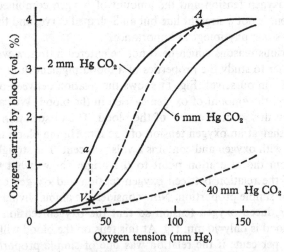

Fig. 12. Oxygen dissociation curves of the blood of *Loligo* in the presence of carbon dioxide. Point '*A*' defines the condition of arterial blood, point '*V*' that of venous blood. The difference between '*a*' and '*V*' shows the extra amount of oxygen which is displaced by the carbon dioxide in the tissues.

per cent on simple proportion; 100 c.c. of blood will thus provide the tissues with $2 \cdot 95 - 0 \cdot 65 = 2 \cdot 3$ c.c. of oxygen as compared with $1 \cdot 95 - 0 \cdot 8 = 1 \cdot 15$ c.c. provided by a system working on simple proportion. The effective oxygen carrying capacity of the blood is increased by reason of the S-shaped dissociation curve. To put it another way, the steeper the dissociation curve over its working range, the more effective is the blood as a transporter of oxygen.

The shape of the curve and its position in relation to the axes are affected by a variety of physico-chemical factors. Increase in temperature or increase in acidity displaces the curve to the right, that is to say these changes reduce the affinity of the blood pigment for oxygen since at any given oxygen tension it will now hold less. The addition of carbon dioxide to the blood increases the acidity and of course this is what happens when the blood passes through the tissues. As the blood picks up carbon dioxide in the tissues and becomes more acid it gives off its oxygen more easily, and as the carbon dioxide leaves the blood in the respiratory organ oxygen is taken up more readily. In *Loligo* the arterial blood contains oxygen at a tension of 120 mm Hg and carbon dioxide at a tension of 2 mm Hg, while for the venous blood the figures are 40 and 6 mm Hg respectively. In Fig. 12 two dissociation curves are shown, one relating to a carbon dioxide tension of 2 mm Hg, the other to a carbon dioxide tension of 6 mm Hg. The point A represents the condition of arterial blood and the point V the condition of venous blood, and in the ordinary course of circulation the blood must follow some curve between A and V, a curve which is steeper than either of the others. Or, alternatively, we may put it that if the oxygen affinity of the blood pigment was independent of the carbon dioxide tension the blood would deliver A–a, that is, $2 \cdot 6$ vols. per cent of oxygen to the tissues; being affected by carbon dioxide it delivers A–$V = 3 \cdot 6$ vols. per cent.

The dependence of oxygen affinity upon carbon dioxide tension is obviously made use of to the animals' advantage in the case we have been considering. But what would happen to *Loligo* if it had to breathe air may be guessed from the dissociation curve of its blood in the presence of 40 mm Hg carbon dioxide, such as obtains in the alveolar air of man. Clearly, the blood would hardly take up any oxygen at all as it passed through the lung. The properties of blood pigments show considerable variation from one species to another and, as one would expect, sensitivity to carbon dioxide is very much less in air-breathing animals than in water-breathing animals; in human blood, for example, an increase of carbon dioxide tension from 3 to 40 mm Hg does no worse than to reduce the oxygen affinity by half. The change from water-breathing to air-breathing thus involves not only the re-design of the respiratory organ but also radical alteration of the properties of the blood pigment.

In earlier discussion of the exchange of gases at the surface and their transport through the body it was convenient to omit all reference to insects, because in terrestrial arthropods having a tracheal system these processes are carried out along entirely different lines. Instead of the usual localized respiratory organ and the transport of gases by the blood we find a system of tubes which penetrate the body and carry air directly to all the organs.

Many of the larger insects ventilate the tracheal system. Large thin-walled air sacs are developed leading off from the main tracheae, and their volume can be altered by contraction of the abdominal muscles. In the locust, for example, air is drawn in through the thoracic spiracles and expelled through the abdominal spiracles and a circulation of air through the main tracheal trunks is maintained in this way.

But in small insects and in the smaller branches of the tracheal system in large insects exchange of gases must be brought about by diffusion alone. In view of the great activity of insects and their high rate of oxygen consumption many physiologists doubted that diffusion alone would be adequate; but about 1920 Krogh made some calculations based upon a large caterpillar which showed no respiratory move ments, and taking into account its rate of oxygen consumption, the rate of diffusion of oxygen and the anatomy of its tracheal system he found that not only could the oxygen supply be maintained by diffu-sion, but that the partial pressure of oxygen at the tracheal endings would be only a few mm Hg less than at the spiracles. The great advantage of the tracheal system is that a high oxygen tension can be maintained in the tissues without energy being wasted in maintaining a rapid flow of blood. The poor development of the heart in insects as compared with crustacea is no doubt related to the fact that in insects the blood has no respiratory function. But on the other hand, since the rate of diffusion is inversely proportional to distance, the tracheal system is not readily adaptable to the needs of larger animals.

Internally the tracheae end blindly in very fine intracellular tubes known as tracheoles (Fig. 13). These are partly filled with fluid, and the extent to which air penetrates down them varies with the state of activity of the insect. If it is very active, or if oxygen is withheld, the air extends further down the tubes and comes nearer to the tissues, thus shortening the final path of slow diffusion in water. One can readily observe this happening because the difference of refractive index between air and water makes the finer air tubes easily visible under the microscope. If a small transparent insect such as a mosquito larva is imprisoned under a cover slip, one can see how as asphyxiation sets in more and more of the finer tubes become visible as the air extends into them.

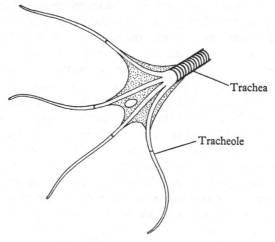

Fig. 13. Diagram of a trachea ending in tracheoles.

We have in the invertebrates many examples of the transition from water-breathing to air-breathing and can often trace the ways in which the original respiratory organ has become modified in the process. The respiratory organs of molluscs, for example, are a pair of gills, known as ctenidia, housed in a special chamber, the mantle cavity, through which water is circulated. In the snail, which breathes air, the ctenidia have been lost and the roof of the mantle cavity has a rich network of blood vessels. The opening of the mantle cavity, which is wide in marine gasteropods, becomes restricted in the snail, and from being a gill chamber the mantle cavity becomes a lung. The insect tracheal system, on the other hand, proclaims itself a respiratory device which has arisen in relation to the terrestrial habit without any evolutionary roots in a pre-existing aquatic organ of respiration; and it is of some interest to study the ways in which the tracheal system has become adapted for water-breathing.

The various aquatic adaptations of the tracheal system, while probably they have been independently acquired, follow two main lines—the closed system and the open system. In the closed system there are no spiracles; exchange of gases with the water takes place through tracheal gills which are thin-walled extensions of the body richly supplied with fine tracheal tubes. They are to be seen in mayfly larvae as dorso-lateral outgrowths on the abdomen, in dragonfly larvae at the posterior end of the body, and in various other insects. In the closed system we have a special aquatic respiratory organ locally developed and the tracheal system retains its role of the transporter of gases about the body, relying of course upon diffusion alone.

The open system shows a wide range of adaptation both in structure and function. A great many aquatic insects are not truly aquatic in that they need to come to the surface to breathe. Mosquito larvae as is well known have their spiracles carried upon a tube, the respiratory siphon, which is thrust through the surface. Other insects, when they come to the surface, take in air stores; the large water beetle *Dytiscus* holds bubbles of air under its elytra and these bubbles are in free communication with the spiracles. Now if the insect carries a bubble of air in contact with the water and in free communication with the spiracles it is in a position to gain oxygen from the water in the following way. As the oxygen in the bubble is used up, the partial pressure of oxygen falls, and if the water is saturated with air there will be a tendency for oxygen to diffuse into the bubble from the water. The bubble thus acts as a physical gill, and a very efficient one, for there is no cuticle or living tissue separating the water from the air. And although water beetles have to come to the surface from time to time to renew their air bubbles, it is undoubtedly true that in well-aerated water they can stay down longer than can be accounted for by the oxygen originally present in the bubble.

If the insect is allowed to take in a bubble of pure oxygen and to dive into water saturated with pure oxygen the result is rather surprising—the insect drowns. When examined it is found that the bubble has disappeared and that there is water in the tracheal system. What happens is this. When the insect takes down a bubble of air into air-saturated water the removal of oxygen from the bubble causes a fall in partial pressure of oxygen because the proportion of oxygen to nitrogen is lowered. Given this difference of partial pressure it becomes possible for oxygen to diffuse in from the water; if this is inadequate and if oxygen is used by the insect faster than it is replaced, the bubble will eventually contain nitrogen only and the insect, suffering respiratory distress, will come to the surface. Nitrogen, therefore, is of prime importance in this mechanism. But when the bubble contains pure oxygen and the water is saturated with oxygen all that happens when the insect uses oxygen is that the volume of the bubble decreases; since it contains only oxygen, the removal of oxygen from it does not alter its composition and there is no difference in partial pressure to cause oxygen to diffuse into the bubble from the water. For the same reasons the insect never experiences any lack of oxygen as it normally would do; it therefore remains submerged while the bubble gradually shrinks in size, and water enters the tracheal system before the insect is aware that anything is wrong.

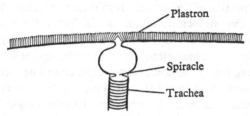

Fig. 14. Diagram of the relation between plastron and tracheal system.

This would not happen in those few insects which possess the respiratory device known as the plastron. This is a pile of very fine non-wettable hairs which cover the cuticle for some distance around the spiracle—Fig. 14. These hairs, holding off the water, maintain a film of air over the surface of the body, and are in communication with the spiracle through grooves which are guarded by the same hair pile. To collapse the hair pile and drive water into the tracheal system a pressure of several atmospheres is required. This device not only extends the surface of the physical gill but also guards the open tracheal sytem against the entry of water. It is probably the neatest and most efficient device for water-breathing evolved in the animal kingdom and it is at the same time perfectly capable of being used in air.

Apart from the insect tracheal system the respiratory arrangements of the lower animals conform to the same general principles as are brought to light in the study of vertebrates. Perhaps this is not so surprising, for the principles are after all very simple ones. A respiratory organ is no more than a part of the body where the internal and external media are separated by a very thin membrane through which gases can diffuse; this process of simple passive diffusion generally has to be supplemented by convection. Simple physico-chemical considerations are very much to the fore in respiratory problems and they are the same for all animals.

The respiratory organ may be the general surface of the body as in the earthworm, or it may be a sharply differentiated structure like the molluscan ctenidium, or it may be a part of some other structure, as in *Nereis* where the thin walls and rich blood supply of the parapodia suggest that in addition to being organs of locomotion they are also organs of respiration. But it is as easy to infer the respiratory significance of an organ from anatomical studies as it is difficult to prove by physiological methods that the organ is responsible for a substantial portion of the animal's respiratory exchange. We are used to the idea that suffocation means death. Yet in the lower animals we can often amputate what we think is the respiratory organ without fatal consequences; the animal often seems somewhat lethargic after the operation, but in due course it will regenerate a new organ and be none the worse. We know that carbon monoxide is a very poisonous gas; it is poisonous because it combines with haemoglobin and displaces the oxygen, thereby putting the haemoglobin out of action as an oxygen carrier. We can likewise use carbon monoxide to put out of action the haemoglobin of invertebrates, but to discover what inconvenience the animal suffers in consequence calls for very careful observations and experiment. This becomes more understandable when we realize what a vast difference there is between the levels of activity sustained by mammals and insects and the levels of activity sustained by many of the lower animals; we can get some measure of this by comparing rates of oxygen consumption.

	Oxygen consumption $mm^3/gm/hr$
Sea anemone	13
Earthworm	60
Octopus	90
Frog	150
Sparrow	6,700
Mouse, resting	2,500
running	20,000
Butterfly, resting	600
flying	100,000

These differences show that in most of the lower animals life goes on at a very much slower tempo. No amount of stimulation could raise their oxygen consumption to the mammalian level, because their respiratory organs and circulatory systems could not get oxygen to the tissues at the mammalian rate. Nor is there any particular reason why they should. Much of what was said at the end of the previous chapter about selection pressure might be repeated here. The snail has its radula with which to feed on plants, plants are everywhere abundant and it does not have to travel far in search of food. When danger threatens it does not try to run away but retires into its shell. The more one looks into the snail's way of life the more one sees that it has nothing to gain by stepping up its activity, and for its low level of activity a respiratory organ and a circulatory system of moderate efficiency are adequate. Natural selection is not interested in physiological efficiency for its own sake but only so far as it can contribute to the effectiveness of feeding or escape or reproduction or any other aspect of its biology that brings the animal up against the rub of the environment.

4

EXCRETION

When we speak of excretion what we generally have in mind is nitrogenous excretion, the elimination of the waste matter of the body. This is certainly the function of the kidney, but not its only function; the kidney is also responsible for maintaining the normal composition of the blood. But while this is true of mammals it is not altogether true of the lower animals. In them we can readily recognize organs which are suggestive of the kidney—the nephridia of earthworms, for example—but such organs are not necessarily concerned in nitrogenous excretion, nor are they the only organs in the body which play a part in the regulation of the composition of the blood. To understand the part played in the body by the so-called excretory organs we shall have to inquire more deeply into the processes of excretion and of the regulation of the salt and water content of the body.

Some of the nitrogenous waste matter excreted from the body is the result of the activity of the repair services, alluded to in Chapter I, but in mammals and probably in other animals as well the bulk of the nitrogenous excretory matter comes directly from the food rather than from the tissues of the body. In a carnivorous animal the food is mainly protein, and after digestion the proteins are absorbed into the body as amino-acids. Most of these amino-acids must be used as a source of energy and the first thing that happens to them is that they are converted into ordinary non-amino organic acids by having their amino groups removed. This process, known as deamination, results in the production of ammonia.

Now ammonia is a poisonous substance, mainly by virtue of its alkalinity, and it must not be allowed to accumulate in the body. But it has a small molecule and diffuses easily through animal tissues. Provided that plenty of water is available there is no more difficulty in getting rid of ammonia than in getting rid of carbon dioxide, and in aquatic animals they often leave the body by the same route. In fresh-water fishes only about 10 per cent of the nitrogenous excretion leaves the body through the kidney; the rest leaves through the general body surface and, one imagines, mostly through the gills. So much for the view that the kidney is the main organ of nitrogenous excretion.

There is no reason why ammonia should not be the main excretory product of an animal provided that sufficient water is available to keep it in dilute solution. But terrestrial animals do not enjoy unlimited supplies of water and cannot afford the relatively enormous amounts required for the excretion of ammonia. In mammals the ammonia is combined with carbon dioxide to form urea. Urea is very much less poisonous than ammonia, but it is a soluble substance of fairly low molecular weight and the osmotic pressure which it would exert in a concentrated solution would be beyond the powers of the excretory organ to overcome. If nitrogenous excretion is to be in the form of urea a considerable supply of water is needed for its elimination, though of course much less than in the case of ammonia.

For terrestrial animals the ideal substance is uric acid. The solubility of uric acid is very low and when water is removed the uric acid precipitates from the solution before the osmotic pressure has risen to a level that matters. It is therefore possible for the animal to get rid of its nitrogenous waste in the form of uric acid with minimum loss of water. Birds, reptiles, and insects adopt this method.

While in aquatic animals ammonia and to some extent urea can be allowed to escape by the general surface of the body, terrestrial animals obviously require special organs of excretion. But if the excretory material is uric acid it is not essential that it should leave the body—it has to be kept out of circulation, that is all. The cockroach can get rid of uric acid through its Malpighian tubules but it also lays it down in the fat body where it simply accumulates during life. In the same way the uric acid produced by the chick is deposited in the allantois and left in the shell when the chick hatches. Organs which store excretory matter in this way are sometimes spoken of as kidneys of accumulation.

It is not altogether easy to set out in a few words the background to the salt and water requirements of animals. The importance of inorganic ions for physiological processes in animals can be traced back through their importance for the physiological processes of cells to their importance for the behaviour of proteins in non-living systems. The cells of all organisms seem to be concerned to maintain within their cytoplasm certain inorganic ions within fairly narrow limits of concentration, by no means the same as the concentrations of these ions in the cell's environment. We can tell, by the use of radioactive isotopes, that the cell membrane is not completely impermeable to these ions; they can leak through it, and if nothing is done about this they would attain the same concentrations inside and outside. But all the time the cell is using energy to 'pump' the ions in or out so as to maintain the differences in concentration. The amount of energy required for this depends, for each kind of ion, upon the ratio of the concentration inside to the concentration outside. The cell therefore cannot be indifferent to the ionic composition of the environment in which it lives.

As natural waters go, sea water is a passable environment for cells to live in. Those ions which the cell requires are present in sea water, and this is probably no coincidence since it seems likely that life originated in the sea. But the ions are not present in the same concentrations as in the cell. Whereas in sea water the principal cation is sodium, in most cells it seems to be potassium; and the concentrations of magnesium and sulphate in cells are much less than in sea water. Cells living in sea water, although they have plenty of ions available, have to use a significant proportion of their available energy to maintain their preferred internal concentrations of ions. A way of avoiding some of this wastage of energy is open to multicellular organisms; it is to set up a special internal environment of their own—the blood. To certain cells of the body is delegated the whole-time task of maintaining the ion concentrations in the blood at suitable levels. This makes things easier for the other cells of the body which, no longer having to fight individual battles with the environment, have more energy available for their other activities.

Provision has also to be made to maintain the water content of the internal medium. For primitive marine animals this is no problem, since cells can readily tolerate a medium which is as concentrated as sea water. But for fresh-water animals a problem does arise. For the sake of its cells the fresh-water animal must have some salts in its blood, and it must therefore develop surface membranes which are relatively impermeable to salts if it is to retain them. But to develop surface membranes impermeable to water seems to be beyond the powers of the lower animals, and the permeability of living membranes to water is always very much greater than their permeability to salts. The fresh-water animal is thus in the awkward position of being an osmometer; water is continually diffusing into it and would cause it to swell and burst if steps were not taken to remove the water. This, as we shall see later, is one of the functions of the excretory organ. The terrestrial animal has the opposite problem, that of preventing all avoidable loss of water, and the excretory organs of terrestrial animals are commonly adapted to this end.

The study of the minute anatomy of the kidney was begun in the middle of the seventeenth century, but it was not until the early part of the nineteenth century that the relations of the tubule to Bowman's capsule and the glomerulus were correctly appreciated. Almost at once theories as to the mechanism of urine formation were put forward. From analyses of urine and on the basis of experiments relating urine flow to blood pressure most physiologists were prepared to accept the idea that the first step in urine formation was a process of filtration from the blood in the glomerulus into the cavity of Bowman's capsule, that this glomerular filtrate contained all the nonprotein constituents of the blood in the same proportions as in the blood plasma, and that in its subsequent passage down the tubule the composition of the filtrate was altered. It was well on in the present century before this view received direct confirmation. In 1925 Richards introduced the technique of pipetting. He was able to insert a fine quartz pipette into the Bowman's capsule of a living amphibian kidney and to withdraw samples of fluid about one cubic millimetre in volume. He and his associates developed methods of analysis suitable for such small quantities and in this way they were able to show that a great many constituents of the urine were present in the glomerular filtrate in the same concentrations as in the blood plasma. The physiologists of the nineteenth century, ready to accept the idea of filtration, were bitterly divided about the function of the tubule. One side maintained that the glomerular filtrate was modified during its passage by absorption of substances from it and by secretion of substances into it; the other side held that all the known facts could be accounted for by absorption alone and that the tubule had no powers of secretion. In 1934 it was found that there are certain fishes which have no glomeruli in their kidneys; this makes it certain that the tubule has powers of secretion, and it is now well known that these powers are exercised in the normal urine production of mammals.

The composition of the urine in vertebrates and its rate of flow are clearly related to the environment in which the animal lives. If a frog is kept in water it suffers an inward diffusion of water through its permeable skin. The water has to be got rid of by the kidneys, but without loss of salts, and these are removed from the glomerular filtrate as it passes down the tubules. A frog produces urine at the rate of 25 per cent of its body weight per day and the urine is hypotonic to the blood (i.e. has a lower osmotic pressure than the blood, owing to the removal of salts). In man there is extensive reabsorption of water as the glomerular filtrate passes through the tubules. Urine is produced at the rate of 2 per cent of the body weight per day and is hypertonic to the blood. In desert-living mammals the amount of urine passed is less and its concentration greater.

Very little is known about the excretory organs of the lower invertebrates. Among the annelids only the earthworm has been studied. The earthworm is generally reckoned a terrestrial animal and it is adapted to terrestrial life to about the same extent as the frog is. But like the frog it has excretory organs which are adapted to getting rid of water rather than to conserving it, and in this respect both the frog

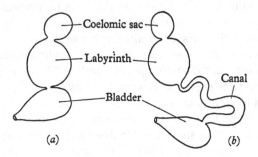

Fig. 15. The antennary gland of (a) *Carcinus* (marine) compared with that of (b) *Astacus* (fresh water).

and the earthworm are fresh-water animals. The nephridia of the earthworm produce a fluid which is hypotonic to the blood at a rate of about 60 per cent of the body weight per day.

In the crustacea we find better examples of variation in structure and function of excretory organs which can be related to modes of life. In the shore crab *Carcinus* the excretory organ is the antennary gland, or green gland, and consists of three parts (see Fig. 15). The coelomic sac opens into the labyrinth, a chamber with folded walls giving the gland its green colour, and this in turn opens into a thin-walled bladder. The urine of the crab is isotonic with the blood. In the antennary gland of the fresh-water crayfish *Astacus* a canal is interposed between labyrinth and bladder. By pipetting and micro-analysis it has been shown that the urine in the coelomic sac and labyrinth is isotonic with the blood and that in the canal salts are reabsorbed with the result that the urine in the bladder is hypotonic to the blood. It must not be supposed, however, that the antennary gland of the crab is no more than a filtering device. The crab's blood is isotonic with sea water but not identical with it in composition; it contains more potassium and less magnesium. The crab's urine is isotonic with the blood but contains much more magnesium and less potassium. It is partly due to the activity of the antennary gland that the difference in composition between the blood and sea water is maintained. To this matter we will return later.

Most text-books designate the Malpighian tubules as the excretory organs of insects, but in fact the role of the Malpighian tubules in excretion is only to be understood when considered in relation to the role of the rectal glands. The fact that the Malpighian tubules open into the hindgut and not directly to the exterior is a matter of great significance. Even if, as is exceptionally the case, uric acid is precipitated while the urine is still within the lumen of the tubule, it would not be practicable to remove all the water from it because some fluid would be needed to flush the crystals out of so narrow a tube. But if the urine is run into the rectum it is then exposed to the powerful action of the rectal glands which are able to remove water from the urine and faeces until only a hard dry mass remains. The faecal pellet can then be voided from the body by the contraction of muscles.

The physiological mechanism of urine formation in Malpighian tubules depends primarily upon the secretion of a solution of potassium chloride into the lumen. This solution, which is of approximately neutral pH, carries with it most of the small organic molecules present in solution in the blood, including the soluble salts of uric acid. On reaching the rectum the urine is acidified, whereupon the uric acid is precipitated. Potassium and organic compounds which are useful to the insect are reabsorbed into the blood along with the water.

The excretory mechanism of the insect appears to have this much in common with the excretory mechanisms of other animals, namely that all soluble substances present in the blood (those of small molecular weight) are allowed to enter the lumen of the excretory organ and those which are of value to the animal are later reabsorbed. Basically, this is filtration-reabsorption, and when one thinks about it one can see that there is much to recommend it. The excretory system of an animal is charged with the task of removing not only the normal excretory products of the animal but also any foreign substances which may have gained access to the body. It would be impossible to provide for every conceivable foreign substance a specific mechanism whereby it could be 'deliberately' excreted into the urine; but by allowing all soluble substances to enter the urine the animal automatically ensures the elimination of unwanted substances merely by not providing specific mechanisms for their reabsorption.

From this survey of the excretory organs and the processes of excretion in animals we are at liberty to draw some general conclusions. In animals which are well adapted to life on land water conservation is a main preoccupation, and in such animals we find that nitrogenous excretion and regulation of the composition of the body fluids are carried out in the same organ. In aquatic animals we find organs which by convention we call excretory, but we do not find, nor should we expect to find, that these organs are the main channel by which nitrogenous waste leaves the body. On the other hand we seldom fail to find some evidence that these organs have a part to play in regulating the composition of the blood and body fluids. In this would appear to lie the primary importance of such organs. The degree of importance attaching to constancy in the composition of the body fluids we will now endeavour to assess.

A remark made by the great French physiologist of the nineteenth century, Claude Bernard, is often quoted today. 'La fixité du milieu intérieur est la condition de la vie libre.' Freely translated, this means that the ability to keep the composition of the blood constant confers the ability to live in a variety of environments. This idea, if it is a sound one, should illuminate much of what we have been discussing under excretion. What steps do animals take to maintain a constant internal environment and what do they stand to gain by it?

When an animal is placed in an unsuitable external medium which causes undesirable changes in the internal medium, as for example when a marine animal is placed in fresh water, there are two ways in which it may react. It may pump out the water as fast as it comes in; this involves the performance of work against the osmotic forces and is called the method of correction. Or the animal may render the body surface less permeable so that less correction is called for; this is the method of evasion. There is nothing mutually exclusive about these methods; they can be, and normally are, pursued together.

To continue with the example of a marine animal placed in fresh water or in dilute sea water. The form of graph shown in Fig. 16 is a very convenient way of expressing the osmotic relations between internal and external media. We express osmotic pressure in terms of freezing-point depression, Δ_i being the freezing-point depression of the internal medium and Δ_o that of the external medium. If we use the same scale for both then a line drawn at 45° through the origin indicates what will happen if the animal has no powers of regulation and $\Delta_i = \Delta_o$ under all conditions.

The large spider crab *Maia* has no powers of regulation and can survive very little concentration or dilution of the sea water in which it lives. *Carcinus*, the shore crab, has no powers of regulation in concentrated sea water but is able to keep up its internal osmotic pressure over a considerable range of dilution. The higher internal osmotic pressure will cause water to diffuse into the crab and this water must be got rid of if the internal osmotic pressure is to be maintained. But

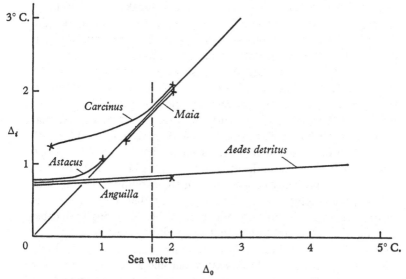

Fig. 16. Osmotic regulation in various animals.
Δ_i = freezing-point depression of blood.
Δ_o = freezing-point depression of external medium.

in this the crab gets no help from its excretory organ which, as we saw earlier, produces urine which is isotonic with the blood. The problem is solved along different lines. There is a continuous inward secretion of salts by the gills which keeps pace with the inward diffusion of water; the antennary gland simply drains off the excess of fluid in the body. Here then is an example to illustrate the remark made in the first paragraph of this chapter, that the excretory organs are not the only organs in the body which play a part in the regulation of the composition of the blood; more examples are to follow.

Maia is confined to the sea. *Carcinus*, by virtue of its ability to regulate the osmotic pressure of its blood, is able to penetrate up estuaries, but it is not able to live in fresh water. *Astacus* normally lives in fresh water and has mastered the problem of maintaining the salt content of its blood. As in *Carcinus*, the gills can secrete salts into the blood and they can do this even when the salts are present in such dilutions as are found in natural fresh waters. The salt-absorption mechanism is supported by the excretory organ which absorbs salt from the urine. But *Astacus* is not able to live in sea water. As the osmotic pressure of the external medium is increased the osmotic pressure of the blood remains fairly constant until the two osmotic pressures are equal; thereafter regulation breaks down and the animal soon dies. For some reason, not yet fully understood, the tissues of fresh-water animals cannot stand the same concentration of salt as can the tissues of their marine relatives.

The ability to absorb salts from fresh water is not confined to the crustacea. It is well developed—and indeed was first discovered—in fresh-water fishes. The osmotic relations of fresh-water fishes are very much like those of *Astacus*; salts are absorbed from the external medium and from the glomerular filtrate. In marine fishes the osmotic pressure of the blood is not the same as that of sea water as one might expect but is about the same as that of the blood in fresh-water fishes; this is probably connected with the evolutionary history of fishes, palaeontological evidence suggesting that they evolved as fresh-water animals and later migrated to the sea. Being permeable to water they are therefore subject to a continuous loss of water from the dilute internal medium to the more concentrated external medium. They replace this by swallowing sea water and absorbing it together with its salts from the gut. The excess salt is then excreted from the body through the gills. Although fresh-water fishes and marine fishes have thus solved their respective osmotic problems very few can pass freely, like the salmon or the eel, from one environment to the other.

The regulatory mechanisms of fishes (meaning the bony fishes, or teleosts) are purely corrective in nature. It would be a pity to leave this subject without mentioning the cartilaginous fishes or elasmobranchs, such as the dogfish. Like the teleosts the elasmobranchs evolved in fresh water and when they migrated to the sea they faced the same problem. This they solved by the surprising method of retaining urea in the blood; the osmotic pressure of the blood of a dogfish is just a little greater than that of sea water and nearly half of it is due to urea, a substance which in most animals the kidney excretes but which the elasmobranch kidney appears to reabsorb from the glomerular filtrate. The difficulty of retaining urea is very much less than the difficulty of fighting the constant drain of water, so that we are justified in classing urea retention as a method of evasion.

Very few invertebrates have achieved the right solution, which is to develop a waterproof skin. The mosquito *Aedes detritus* breeds in salt marshes where the pools may be diluted with rain water or highly concentrated by evaporation. But having an impermeable cuticle the larva is able to evade these changes. It is exposed to the external medium only in so far as it swallows water with its food; the very limited osmotic stress which this imposes can be dealt with by the glands of the rectum.

An animal which lives and has lived only in the sea has no need of any means of osmotic regulation. Marine animals can penetrate up estuaries by evolving the mechanism of inward secretion of salts, but if they are to live permanently in fresh water their excretory organs usually cooperate by producing salt-free urine. The blood of fresh-water animals is generally more dilute than that of marine animals ($\Delta_i = 0.7°$ C as compared with $1.86°$ C) and as mentioned above a fresh-water animal cannot tolerate a salt concentration in its blood such as is normal for a marine animal. In consequence of this, fresh-water animals returning to the sea develop mechanisms for keeping down the salt concentration of the blood. Very few animals have at their disposal corrective mechanisms which enable them to pass freely between salt water and fresh water. Those that can are for the most part animals with impermeable skins—insects, reptiles, birds, mammals—which have followed the method of evasion rather than the method of correction.

Successful adaptation to terrestrial life also demands the development of a water-tight skin, as is found in reptiles, birds and mammals and in the insects. Amphibia, snails and earthworms, with moist skins, make a poor showing as terrestrial animals and can only venture abroad in cool moist air. Even in the best adapted animals some evaporation of water from the respiratory surfaces is unavoidable but it can be kept to a minimum. Most insects have devices for closing the spiracles which are normally kept open only just as much as is necessary to meet the respiratory needs of the moment.

In the field of respiration we have no evidence that animals can secrete respiratory gases against concentration gradients and this means that shortage of oxygen in the environment cannot be overcome by correction. But the possession of a blood pigment with a high affinity for oxygen, so often found in animals living under oxygen deficiency, is a means of getting more oxygen to the tissues and is to be classed as a method of evasion.

Temperature control is not found outside the mammals and birds, for it depends for success upon the development of an insulating layer in the skin—hair, feathers, or blubber. This is obviously the evasion of heat loss. Shivering on the other hand is a corrective mechanism, a means of increasing the heat production in the muscles.

There is any amount of evidence, which it would be tedious to specify further, that animals have evolved methods of making themselves independent of variations in the external environment. This is called homeostasis. Is it rewarded by 'free life'?

Let us ask first what has happened to those animals which have failed to secure this independence. Of these we could hardly hope for a better example than the echinoderms. An animal which makes a practice of pumping the external medium through its body obviously starts with certain disadvantages. The echinoderms have virtually no progress to report; their body fluids are practically identical with sea water, not only in osmotic pressure but in the relative concentrations of the ions present, and all changes in the external medium are faithfully followed. Echinoderms are restricted to the sea; but it must be admitted that they do very well there, in spite of their crazy organization.

Mammals and birds have achieved constancy to a degree unapproached by other animals. They are found in all environments. A mammal can live on land, it can also go for a swim in the sea or in a fresh-water lake. To this extent it is free in a way that the echinoderm is not. Yet this freedom is limited in other ways. Constancy of the internal medium may enable one to keep alive in an environment but it does not mean that one can make a living there. Polar bears, by virtue of their powers of heat conservation, can live in the Arctic; they could also live in temperate regions but they do not do so for the very obvious reason that they could not catch the sort of animals that live there. What free life means is that when an animal has evolved a means of maintaining the temperature of its body in cold air it is then in a position to undergo further evolutionary change and find itself an ecological niche in a cold climate. Mammals are in this position, reptiles are not. Free life is therefore to be thought of in an evolutionary or racial sense rather than in the sense of individual freedom.

In man the degree of constancy achieved is very high. For the body temperature in health the limits are $\pm 0.05°$ C. If the blood shows a tendency to become acid the respiratory centre is stimulated and by increased ventilation of the lungs the removal of carbon dioxide is speeded up. The pH of the blood does not vary by more than ± 0.05.

In muscles, nerves and other tissues which we can study as physiological preparations, variation of performance between these limits of temperature and pH would be barely detectable. How then are we to account for this extraordinarily accurate regulation on the basis of natural selection? The answer appears to be that it is necessary for the higher functions of the brain. If the composition of the blood is changed in any way it is in the brain that symptoms first appear. It is man's brain that has raised him to the dominant position he now holds among animals and by which he has gained the freedom to establish himself where he will upon the earth. For man at least the relation between 'fixité' and 'vie libre' is abundantly demonstrable.

5

CHEMICAL COORDINATION

This is perhaps an opportune moment to introduce the idea that the activities of animals may be assigned to one or other of two categories: maintenance activities and operational activities.

To take the second, operational activities include all the overt activities of the animal which are directed towards obtaining its food, escaping its enemies and finding a mate. These activities are brought into effect by the contraction of muscles and the muscles are made to contract by nerves. The animal becomes aware of its surroundings through its sense organs and the sense organs are functionally related to the nerves and muscles through the central nervous system. The nervous system is responsible for ensuring that the contractions of the muscles are appropriate to the animal's environmental situation. This is nervous coordination, which we shall take up later, in Chapter 9.

The maintenance activities of an animal are less obvious to the casual observer, but they are going on all the time, even when the animal is asleep. Maintenance activities include digestion, excretion, respiration and circulation—in fact, just those activities with which we have been concerned in the foregoing chapters. Like operational activities, maintenance activities have to be coordinated, and in some cases, notably respiration and circulation, the nervous system is responsible for this. These cases, however, are the exceptions. For example, in the mammalian body the composition of the blood is regulated by the kidney. If the animal drinks a large volume of water, this is soon got rid of as a large volume of dilute urine; if the animal is deprived of water, the water content of the body is conserved by the excretion of only a small volume of urine which is correspondingly more concentrated. Although the kidney has a nerve supply it has not been possible to discover any effect of stimulating these nerves upon the composition of the urine, and the adaptive responses of the kidney to the state of water balance in the body are not abolished when the nerves are cut. The control of these adaptive responses is exercised through a hormone circulating in the blood. There is a centre in the hypothalamus at the base of the brain which is sensitive to the concentration of the blood; when the concentration rises above its normal value this centre causes the pituitary gland to release into the blood an anti-diuretic hormone which promotes the re-absorption of water from the glomerular filtrate. This is an example of chemical coordination.

The mechanism of chemical coordination was first discovered by Bayliss and Starling at the turn of the century. They had set out to study the means whereby the secretion of the alkaline pancreatic juice was timed to coincide with the arrival of the acid contents of the stomach in the duodenum. The cutting of the nerve supply to the pancreas did not abolish the effect, and a secretion was still evoked even when acid was injected into a short loop of small intestine whose nervous connections with the rest of the body had been severed. These observations suggested that the channel of communication between the intestine and the pancreas must be the blood stream. Bayliss and Starling went on to show that some substance with excitatory effect upon the pancreas could be extracted from the intestinal mucosa by treatment with acid; they named this substance 'secretin'.

Following this discovery Starling, with prophetic insight, foresaw the possibility that many more physiological functions might be controlled in this way, and he proposed the name 'hormone' for a class of substances which had hitherto been called 'chemical messengers'.

The student who has dissected the rabbit or the rat will remember that he was asked to observe the thyroid gland and the adrenal glands, and possibly also the pituitary gland at the base of the brain. These organs are all anatomically conspicuous, and for many years now we have known that they are endocrine glands (or ductless glands, or glands of internal secretion) whose function it is to secrete hormones into the blood stream. Other tissues of known endocrine function are not found as discrete organs. The hormone insulin is secreted by certain cells in the pancreas which are present in small groups, known as the islets of Langerhans, and can be seen only in sections; and the sex hormones are secreted by the interstitial cells of the ovary and testis, which likewise cannot be recognized except in sections. These anatomical differences have been of some consequence in the history of physiological research. Where the gland is a discrete body and is surgically accessible the investigator is obviously invited to remove it and see whether anything goes wrong; and if it does he may then make an extract of the gland and inject it into the operated animal in the hope that it will relieve the symptoms. But this line of approach cannot be followed in the case of endocrine tissues which like the islets of Langerhans, are embedded in organs of other function, and indeed may have escaped the notice of anatomists altogether. The physiological significance of these hidden endocrine tissues has usually been discovered by a careful study of unexpected consequences following upon the surgical removal of the host organ. It was in this way, too, that the endocrine systems of the lower animals were brought to light.

There would be no point in attempting to present a comprehensive account, even in barest outline, of what is known about the mammalian system of chemical coordination. But in as much as we have taken the mammal as a standard of comparison in the case of other physiological systems, it is appropriate that at least we should examine the principles upon which the mammalian system operates. As an example, then, we will take the hormonal control of reproductive activities.

The onset of sexual maturity in a mammal is signalled by the increased secretion of certain hormones from the anterior pituitary (adenohypophysis). These are known as gonadotrophic hormones because they promote the growth of the ovaries and testes. The ovaries and testes are so-called target organs of the gonadotrophins. In their turn the fully developed ovaries and testes secrete increased amounts of sex hormones, respectively oestrogens and androgens, and the sex hormones have as their target organs other structures in which are manifested the secondary sexual characters. These include the genital tracts and their glands in both sexes, the mammary glands in the female, the hair follicles of the face in the human male, and so on. Another effect of the sex hormones is that as their concentration in the blood rises they depress the production of gonadotrophins by the pituitary. By analogy with the devices of the electronic engineer this relationship between the gonads and the pituitary is commonly described as negative feed-back; it serves to stabilize the concentration of sex hormone in the blood. A feed-back mechanism of the same type has been shown to operate in the cases of the thyroid and the adrenal cortex which are also under the control of the pituitary. In mammals the pituitary, which secretes a great variety of hormones, is in a position to control and coordinate the levels of activity in other endocrine glands.

Mammals of the primate order are able to breed throughout the year, but in many mammals reproductive activity is restricted to a breeding season. The breeding season is usually the spring or early summer, and it has been shown in many cases that the significant environmental factor which determines the onset of breeding condition is light—the gradual increase of day length. Although the precise mechanism of the effect of day length has yet to be worked out, it is clear that the operative pathway is: eyes → hypothalamus → adeno-hypophysis → gonadotrophins → sex hormones. In this way, mainly by chemical coordination, breeding activity is brought into line with seasonal changes in climate, so that the young are born at a time when they have best chance of survival. Comparable arrangements for the chemical coordination of breeding activity have been described in other classes of vertebrates.

It is obvious, too, that successful reproductive activity must involve close cooperation between chemical coordination, which prepares the reproductive organs for their function, and nervous coordination, which by bringing the sexes together in the act of copulation enables the purposes of these organs to be realized. The sexually appetitive behaviour, which is manifested in the breeding season, can be experimentally evoked at other times by injection of the appropriate sex hormone. In this way the chemical system has its influence upon the nervous system. Conversely, through the close connexion between the hypothalamus and the pituitary the nervous system can make itself felt upon the chemical system. In the rabbit, for example, ovulation does not take place until after copulation, the stimuli received by the genital organs of the female being transmitted by nerves to the brain and thence by chemical message from the adenohypophysis to the ovary.

At the histological level the precise anatomical relations between the hypothalamus and the adenohypophysis are imperfectly known in mammals; but in the lower vertebrates, and in mammals in the case of the posterior pituitary (neurohypophysis), the relations are much clearer. There is in the hypothalamus a group of nerve cells whose axons run to the neurohypophysis. These cells are recognizable in sections by the presence in their cytoplasm of granules having distinctive staining properties; and circumstantial evidence suggests that these granules are manufactured in the body of the cell, passed along the tubular axon to the neurohypophysis and stored there pending release into the blood stream. Cells of this type are called neurosecretory.

In a later chapter we shall see that in the case of motor nerves to muscles there is also evidence of secretory function. The nerve impulses which pass very rapidly along the nerve fibre are to be understood in terms of the movements of inorganic ions across the surface of the fibre, as will be explained in Chapter 7. When an impulse

63

reaches the neuromuscular junction a minute amount of an organic ion, acetylcholine, is secreted by the nerve ending into the space between it and the muscle fibre and excites the muscle fibre to contract. Neurosecretory cells are not perhaps so entirely different from ordinary nerve cells as their appearance and activities might suggest. It could even be that the method of translocating chemical messages bodily along a tube is the more primitive form of nervous action, and that the nerve impulse is a device which has been evolved to avoid the delays of translocation over long distances—but this is pure speculation. Be that as it may, it is worth noting that among the lower animals neurosecretory cells figure prominently in their systems of chemical coordination; it is endocrine tissue of non-nervous origin which is less frequently encountered than in mammals.

Let us now look at a few examples of chemical coordination from the invertebrates. For coelenterates, platyhelminthes, nematodes and echinoderms we have no information at all. In annelids the regeneration of amputated segments has been shown to be under the control of neurosecretory cells in the supraoesophageal ganglion. In *Nereis* the posterior few segments can be regenerated; regeneration fails if the worm is decapitated at the same time, but does take place if one implants into the decapitated worm the brain of another worm from which the posterior segments had been amputated three days previously. These observations implicate the brain in the control of regeneration, and it seems likely that the neurosecretory cells liberate into the blood stream some hormone which activates growth. Another effect of the removal of the brain in annelids is to bring on sexual maturity, and this can be prevented by implanting the brain of an immature animal. In this case it would seem that the brain of a young animal produces a hormone which prevents the development of the reproductive organs. In cephalopod molluscs there are two small glands, one on each side, upon the optic stalks and these produce a hormone which encourages the development of the reproductive organs. No examples of endocrine glands, even of neurosecretion, have yet been found in the other classes of molluscs. By far the greatest amount of information that we have comes from the arthropods, in fact from the crustacea and the insects. The phenomena which have been shown to be under chemical control are principally moulting, metamorphosis, yolk deposition, water balance and colour change.

In many animals colour changes are brought about by cells in the skin, known as chromatophores. These cells are relatively large and of irregular shape, and they contain granules of pigment. When the granules are gathered together into one part of the cell their colour is not conspicuous; when they move so as to distribute themselves throughout the chromatophore the colour becomes visible. Many years ago it was observed that the extirpation of the eye stalks in crustacea led to changes in colour, and it is now known that within the eye stalk there is quite a complicated system of endocrine glands and tissues. The first structure to which endocrine activity could be traced was a small vesicle to which the name sinus gland was given. It was found to be well supplied with nerves, but there was little about the structure of its own cells to suggest that it was the source of some hormone. At that time—between the two world wars—the widespread occurrence of neurosecretion was as yet unrecognized, and it was not until about twenty years later that the origin of the hormones was traced to neurosecretory cells within the nervous system. The sinus gland, then, turns out to be not an endocrine gland in the true sense but what is now called a neurohaemal organ, in which the products of neurosecretion are temporarily stored before their liberation into the blood stream. We have already noted this function in the mammalian neurohypophysis.

This example, it must be understood, does not encompass all that is known about chemical coordination in crustacea, but in this brief treatment of the subject we shall find it more profitable to turn to the insects which have been more intensively studied and whose endocrine mechanisms are more fully understood. One reason for this is the remarkable ability of insects to stand up to surgery. Some insects can remain alive for more than a year after decapitation. It is also possible to unite individuals in parabiosis, whereby their blood streams are made confluent, and this is a powerful technique for the study of blood-borne substances which can be shown to be released in one individual and to take effect in the other.

The process of development in insects, as in all arthropods, is punctuated by a series of moults, or ecdyses. Moulting is under hormonal control, and the principal features of the endocrine mechanism have been worked out. The structures involved are illustrated in Fig. 17. Just behind the brain, dorsal to the oesophagus and in close association with the dorsal aorta, are two pairs of small bodies, the corpora cardiaca and the corpora allata. They are connected by nerves to the brain and to other parts of the nervous system. Further back in the body, in the prothorax, is another gland, often diffuse and of irregular shape, known as the ecdysial gland (or prothoracic gland).

Moulting is ultimately controlled by a small group of neurosecretory cells in the brain. The axons of these cells run to the corpora cardiaca, which are neurohaemal organs, and from here the neurosecretion known as the ecdysiotrophic hormone (or brain hormone) is liberated into the blood stream. On reaching the ecdysial gland it evokes the secretion of ecdysone (or prothoracic hormone) into the blood stream, and it is this hormone which initiates the epidermal changes resulting in the moult.

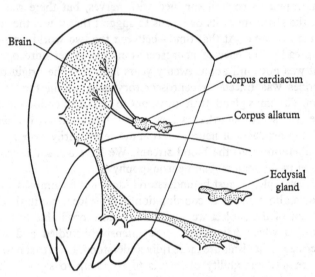

Fig. 17. Head of an insect, to show the endocrine organs.

This particular example has been chosen because it illustrates how in the insect, as well as in the mammal, a hierarchical system of chemical coordination has been evolved, one hormone serving to evoke the production of another. What is found in the mammal, and so far has not been described in the insect, is negative feed-back from the subordinate component to the controlling component of the system.

At the chemical level the preparatory work of extracting and purifying hormones, which are present in exceedingly small amounts, is infinitely laborious. For example, to provide 750 mg of crystalline ecdysone four tons of silkworms were needed. Nevertheless considerable progress has been made in working out the chemical structure of hormones, certainly in mammals. The hormones of the pituitary gland are peptides, but not all of them have been fully worked out. The essential structure of thyroxine and adrenaline have been know for some time. The sex hormones and the hormones of the adrenal cortex—of which there are a great many—belong to a class of cyclic compounds known as steroids. To this class also belongs the insect hormone ecdysone.

At the biochemical level progress is less satisfactory than might have been hoped. Although we may be able to identify the general field of metabolic activity in which a hormone exerts its influence, the biochemical details in that field are often as yet unknown. There is evidence that hormones influence the permeability of membranes, the activity of enzymes, even the activity of genes. But we are in no position to suggest how their influence may be understood in terms of their chemical structure.

At the physiological level information accrues rapidly, but becomes increasingly difficult to interpret. In fact, we have already just about exhausted the capacity of this subject for generalization. Within one species of animal a given hormone may have more than one target organ; a given organ may be the target for two or more hormones; two hormones which act synergically upon one target organ may act as antagonists upon another. When we compare one species with another we find that while the same hormone may be present in both its effects are different; extracts of eye stalk which concentrate the pigments of one species of crustacean are found to disperse the pigments of another. It never seems to be possible to say the last word about anything.

As is true of most physiological activities, we know more about chemical coordination in the mammal than in any other type of animal. From what we already know of insects we can be sure that they too make great use of chemical coordination. But as we go down the scale of beings we find that there is less and less evidence for chemical coordination. Are we then to suppose that chemical coordination is something which is more or less unknown in the lower forms of life and which has been evolved and perfected only in relation to the higher types of organization?

Nothing could be further from the truth.

If we place the widest possible meaning upon the term chemical coordination, extending it to cover all cases in which a substance produced by one cell can exert a controlling influence upon the activities of another, we find abundant evidence of chemical control in all organisms from the highest to the lowest. The discovery of penicillin stemmed from the observation that the growth of the fungus *Penicillium* upon a plate culture of bacteria—the result of accidental infection by air-borne spores—brought about the death of the bacteria in the immediate vicinity. It is now known that penicillin is secreted by the fungus into the medium in which it lives, and that it acts by preventing the growth of the bacterial cell wall so that the bacterium dies when it attempts to divide. The clinical importance of this discovery has distracted attention from its general biological significance, but it did in fact draw aside the curtain to reveal the efficacy of chemical weapons in the fiercely competitive world of micro-organisms.

One kind of micro-organism can secrete into the environment a substance which has a highly specific effect upon another. Such effects are not invariably harmful. Continuing with the fungi we find examples where the hyphae of one mating type produce a secretion which causes hyphae of the other mating type to grow towards them and thus to promote sexual reproduction. In plants the cells of the apical shoot produce substances which, on translocation to other shoots, inhibit their growth and so ensure apical dominance. Over many years evidence has been accumulating to show that in multicellular organisms the differentiation of their cells and the characteristic shapes into which they grow are controlled by substances which are produced locally and diffuse away from the centres of production, thereby setting up gradients of concentration which dictate the patterns of growth.

The production of chemical messengers, far from being a feature exclusive to the higher animals, is to be seen as a universal attribute of cellular organisms. It would not be unrealistic to take the view that all cells, whether they are unicellular organisms or make up the tissues of multicellular organisms, are able to elaborate and secrete substances which can influence other cells, and that they do this as it were in their spare time, while pursuing their more obvious activities. Looked at from this angle we see the endocrine glands of mammals as arising by the specialization of certain cells of the body for these side-line activities, and their assembly into organs which are anatomically recognizable.

6

MUSCLE

In the bodies of animals there are various effector organs—cilia, nematocysts, and the like—which 'do' things in the mechanical sense, but by and large the movements and familiar activities of animals are brought about by muscles. To appreciate the ways in which muscles act it is necessary to consider not only the nature of

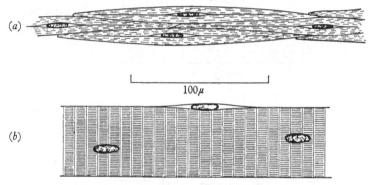

Fig. 18. (a) Unstriated, and (b) striated muscle.

muscular tissue but also the relation of the muscles to the skeleton, through which their movements are transmitted, and the relation of the muscles to the nerves, by which their activity is controlled.

That muscles are either striated or unstriated is a classical proposition of histology. The muscles associated with the skeleton in mammals when examined under the microscope exhibit transverse striations which are not seen in the muscles associated with the viscera. Unstriated muscles have a cellular structure similar to the cellular structure of most other tissues: each muscle fibre is a single cell containing a single nucleus. In the stretched condition its length is of the order of millimetres. In striated muscle the muscle fibre is often some centimetres in length. It contains several nuclei which are not separated from one another by cell membranes. These distinctions between striated muscle and unstriated muscle are not absolute. The heart muscle of mammals has relatively short uninucleate fibres with faint cross-striations. And when we contemplate the varied histological picture presented by muscular tissue throughout the animal kingdom, in which almost every conceivable intergradation of the two types can be seen, it becomes clear that the differences are not fundamental but rather of degree.

The electronmicroscope has now revealed the fine-structural basis of the striations, and from a knowledge of the fine structure it has at last been possible to advance a theory of the mechanisms of muscular contraction which has commanded general support.

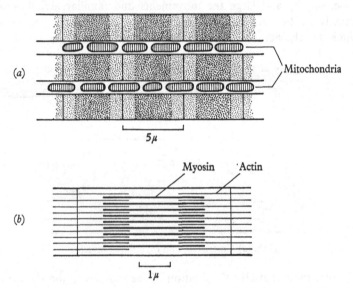

Fig. 19. (a) Three fibrils, with mitochondria between. (b) A single sarcomere, showing filaments of actin and myosin.

The muscle fibre is made up of a large number of fibrils. Like the fibre itself, the fibrils are striated and the striated appearance of the fibre depends upon the alignment of the fibrils with their striations perfectly in register. The spaces between the fibrils are occupied by rows of mitochondria. In their turn the fibrils can be resolved into filaments, which are arranged as shown in Fig. 19.

Each fibril is divided up by transverse membranes into short lengths called sarcomeres, corresponding to the striations. There are two kinds of filament: thin filaments (made of a protein called actin) which are attached to the transverse membranes on either side but fail to reach as far as the middle of the sarcomere; and thick filaments (made of a protein called myosin) which occupy the middle of the

MUSCLE

sarcomere but do not extend as far as the membranes. Over most of
the sarcomere the filaments overlap, and in a transverse section of
this region it is seen that there is a regular hexagonal arrangement,
each myosin filament being surrounded by six actin filaments. When
the muscle contracts the filaments slide over one another and the
sarcomere shortens, not by any shortening of the filaments but by
the greater extent of their overlap.

Unstriated muscles are also made up of filaments of actin and
myosin, but the arrangement of the filaments is less regular and
transverse membranes are lacking.

Actin and myosin can be separately dissolved out of muscle and
precipitated in the form of fine fibres. Actin and myosin together in
the same solution combine to form a complex, actomyosin, and this
also can be precipitated as fine fibres. Fibres of actomyosin, but not
fibres of actin or of myosin separately, will contract in the presence
of ATP.

Thanks to the electronmicroscope, great progress has been made
in elucidating the fine-structural basis of shortening. Biochemists
have discovered how and in what form the energy is supplied. But
the means whereby the energy of ATP is used to cause the actin and
myosin filaments to slide over one another remains obscure and
controversial.

The distribution of striated and unstriated muscles in the body
becomes easier to understand when we realize that the difference in
structure is correlated with a difference in mechanical properties.
Briefly, quick-acting muscles such as those on which we depend for
movement are striated, slow-acting muscles such as those which move
the food along the alimentary canal are unstriated. This correlation
applies not only to the skeletal and visceral muscles of mammals. It
applies also to the contractile cells of coelenterates. In the jellyfishes,
where the contraction of the bell is twitch-like, the contractile elements exhibit traces of cross-striation; no such cross-striation is seen
in the more slowly acting muscles of the sea anemones.

The speed of a muscle is conveniently measured by attaching it to
a device which will record changes in tension and then applying a
brief stimulus to it. The time required to develop maximum tension
is taken as a measure of the speed of contraction and as a measure
of the speed of relaxation we take the time for the tension to fall to
half its maximum value, since owing to its gradual decline it is
difficult to decide exactly at what point the curve reaches the base
line. (Fig. 20.)

71

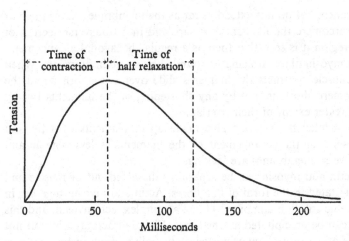

Fig. 20. Tension-time curve for mammalian muscle.

Some figures showing speeds of contraction and relaxation for various muscles are given in the accompanying table.

		Time of contraction ms	Time of half relaxation ms
Cat gastrocnemius	striated	39	40
Cat uterus	unstriated	180	270
Insect leg muscle	striated	35	40
Snail retractor	unstriated	300	3,000
Pecten fast adductor	striated	100	—
slow adductor	unstriated	500	45,000

Pecten has been included in the table because it illustrates very neatly the way in which one and the same movement is brought about either rapidly by a striated muscle or slowly by an unstriated muscle. Lamellibranch molluscs have either one or two adductor muscles which hold the valves of the shell together. These muscles are unstriated and are noteworthy for their very slow rate of relaxation. By reason of this they are well suited for keeping the valves closed over long periods with a minimum expenditure of energy. *Pecten* is unusual among lamellibranchs in being able to swim by flapping the valves of its shell. It has a single adductor muscle which on examination proves to have two components—a striated portion which is used in swimming and an unstriated portion which is used to keep the valves closed.

The relation of the muscles to the skeleton is, as we might expect, bound up with the mechanical properties of both structures. Many of the lower animals do not have a skeleton in the accepted sense. We tend to think of a skeleton as something hard, like the bones of vertebrates or the cuticle of arthropods, but if we set a wider meaning upon the word and regard as a skeleton any device which is capable of transmitting forces from one part of the body to another, we will have to admit that the body fluids of the lower animals are skeletons. The use of a hydrostatic skeleton of this type is shown nowhere better than in the earthworm. The main musculature of the earthworm consists of the longitudinal and circular muscles of the body wall. Elongation of the body is brought about by the contraction of the circular muscles, the tension which they develop being transmitted through the coelomic fluid to stretch the longitudinal muscles. This use of circular and longitudinal muscles in conjunction with a hydrostatic skeleton is the mechanical basis of movement in coelenterates such as *Hydra*, in platyhelminths, in annelids and in many molluscs. In the higher vertebrates layers of longitudinal and circular muscle envelop the intestines; in this case, of course, it is the fluid contents which are moved rather than the intestines themselves. The principle, however, is the same and this was recognized by Jordan who spoke of 'hohlorganartige Tiere'—animals having the form of hollow organs. It is relevant to our present discussion to point out that such animals and such organs are almost invariably operated by unstriated muscles.

On the whole the movements of these animals are slow. The fastest moving animals are to be found among the arthropods and the vertebrates which make use of rigid jointed skeletons and striated muscles. In their limbs the muscles are usually attached close to the joints across which they work, and the fact is that for speed and efficiency there are certain advantages in operating the muscles under conditions of mechanical disadvantage. For reasons connected with the intimate nature of the contractile process muscles work more efficiently when fairly heavily loaded. Muscle has considerable viscosity and if it shortens rapidly an appreciable amount of energy is lost in overcoming the viscous resistance of the muscle substance; this loss is reduced by reducing the distance through which the muscle is allowed to shorten. In the leg of a horse we see how the great muscles are gathered together at the base of the limb and merged into the contours of the body while the distal part of the limb is reduced to a system of jointed rods operated by tendons. If it were necessary for the muscles to undergo considerable shortening in order to develop their power they would have to be attached farther

73

down the leg; the leg would then have a greater moment of inertia and so the horse's speed would be reduced.

In our own bodies the arrangement with which we are familiar is that a muscle has its origin upon one bone—that is, the muscle fibres are attached to the bone—and at the other end it is prolonged into a tendon which is inserted into another bone. The mechanics of the arthropod skeleton and muscles are less widely known. It is commonly supposed that the rigidity of the arthropod skeleton is due to chitin. This is by no means true; pure chitin is a papery substance with no great strength. In the cuticle of the arthropod the chitin is stiffened by other substances. Calcium carbonate is laid down in the chitin of the crustacea, and in the insects the chitin is impregnated with a protein which is rendered very hard by a process of tanning. These hardening processes do not take place equally over the whole body. In the regions of the intersegmental membranes of the body and the arthrodial membranes of the limbs the cuticle remains thin and flexible, permitting movement.

The muscles are attached to the hardened regions of the cuticle and where the muscles are particularly powerful we find that the cuticle is often folded inwards, giving a greater surface for attachment. In the body it is not uncommon for the muscles to bridge across the intersegmental membranes, but in the limbs it is more usual for the muscle to be confined to one podomere[1] and to operate the joint by means of an apodeme, which is the arthropodan analogue of a tendon. The connexions are shown in Fig. 21 b. The apodeme is a long thin invagination running from the hardened cuticle of one podomere into the interior of the next podomere, where muscle fibres connect it to the cuticle. This arrangement is clearly governed by the same mechanical principles as apply to the limbs of mammals.

An entirely different principle is involved in the movement of the wings in insects. The main muscles which cause the up and down movements of the wings are not attached to the wings at all. They are attached to the walls of the thorax and by pulling on the walls they alter the shape of the thorax. It is the alterations in the shape of the thorax that cause the wings to move. Fig. 22 shows how this comes about.

The wing is attached both to the tergum and to the pleuron. The pleural attachment is outside, i.e. lateral to, the tergal attachment

---

[1] Unfortunately the word 'joint' is commonly used not only of the place where the bending occurs, but also of the rigid part between two such places; for the latter the word 'podomere' will be used here.

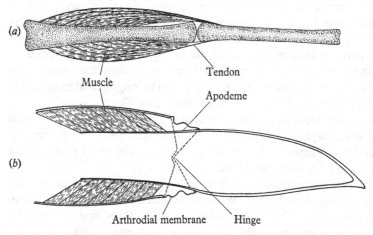

Fig. 21. Relations of muscle to skeleton in vertebrates
as compared with arthropods.

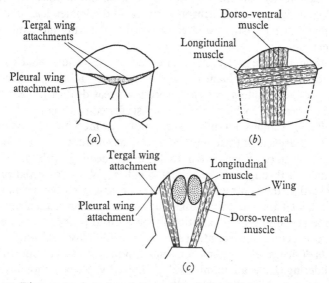

Fig. 22. Diagram to show the relations of the indirect flight muscles of insects.
(a) Thoracic segment seen from the left. (b) Left half of segment removed, to
show the muscles in the right half. (c) Transverse section through thoracic
segment.

and therefore if the tergum moves up relative to the pleuron the wing will move down. The tergum is curved and the logitudinal muscles form, as it were, the chord of the arc. When they contract the curvature of the tergum is increased, the tergal wing attachment moves up and the wing moves down. The dorso-ventral muscles running from tergum to sternum act in opposition to the longitudinal muscles and raise the wing. Thus the thorax works rather like an oil can, and this arrangement is an extreme case of muscles working under mechanical disadvantage. The movements of the thorax are barely perceptible and the distance moved by the wing-tip must be many times greater than the change in length of the muscles.

These few examples which we have chosen serve to illustrate some of the ways in which muscle has become adapted in relation to special requirements. We may compare the rapidly responding limb muscles of vertebrates and arthropods with the adductor muscles of lamellibranch molluscs which have to remain contracted for hours on end. They also show how Nature is quick to exploit the properties of the materials at her disposal, as witness the local development of flexibility and rigidity of the arthropod cuticle in the jointed legs, and the use made of its elasticity and compliance in the wing mechanism of insects.

Although at the molecular level we do not yet know what causes the actin and myosin filaments to slide over one another or how the energy of ATP enters into this process, a great deal of information has been gathered which provides a basis for comparing the muscles of different animals not only in respect of their mechanical properties but in all aspects of their physiology. What emerges from this comparison is a clear indication that we are dealing with the same mechanism throughout the animal kingdom. We have no reason to believe that the contractile mechanism of, say, a molluscan muscle is different from the contractile mechanism of mammalian muscle. Although there is great variation in mechanical properties and even in fine structure, this variation shows no striking discontinuities. The effects of drugs, of inorganic ions and of other agents, although often bewildering, have an unmistakable similarity wherever we investigate them. It appears, too, as if other types of movement, such as amoeboid movement and the movement of cilia, are of the same nature as muscular contraction, so that the ability of muscle, *par excellence*, to contract probably represents the intensive specialization and development of a property which resides in all cells.

7

NERVE

A nerve as one sees it in dissection is properly called a nerve trunk. It is made up of many nerve fibres, both motor nerve fibres running to the muscles and sensory nerve fibres running from sense organs to the central nervous system. These nerve fibres are the extensions of

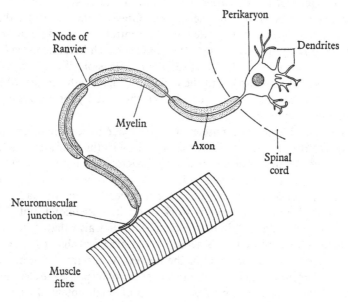

Fig. 23. Diagram of a vertebrate motor neurone.

nerve cells, or neurones. Fig. 23 is a diagram which will serve to identify the principal features of a motor neurone in a mammal. The cell body, that part surrounding the nucleus and for this reason called the perikaryon, lies in the spinal cord. It is of irregular shape, being extended into fine branches known as dendrites through which it makes contact with other neurones; these small areas of contact are called synapses. From one part of the cell body there grows out a more conspicuous extension, the axon, which is the correct word for what we loosely describe as a nerve fibre. The axon may be only about 2μ in diameter, but by the standards of cellular dimensions it can be of enormous length since it extends all the way to the muscle, a distance which may be of the order of one metre. Around the axon is wrapped a sheath of myelin, a fatty material with insulating properties, interrupted at intervals of about one millimetre by the nodes of Ranvier. During its course from the spinal cord to the muscle the axon divides several times and eventually comes to have something like 150 branches, each of which ends on a muscle fibre. At the neuro-muscular junction the axon runs along the membrane of the muscle fibre, indenting it slightly, but separated from it by a narrow cleft.

The fact that muscles are made to contract through the agency of the nerves was known to the ancient Greeks. They believed that 'animal spirits' were generated in the central nervous system and flowed through the nerves into the muscles which became distended and shortened in consequence of their distension. This view and variants of it were held until the end of the eighteenth century when Galvani's experiments suggested that the activity of nerves was electrical in nature.

Galvani showed that muscles could be made to contract if electric shocks were applied to their nerves, and the scientific study of nervous phenomena dates from this discovery. Progress was not very rapid at first because electricity was still a relatively new phenomenon to physical science. By the middle of the nineteenth century it had been established, principally by the work of du Bois Reymond, not only that nerves could be stimulated electrically but also that if natural stimuli were applied to the sense organs electrical disturbances of some sort could be detected in the nerves. As the recording instruments were improved in sensitivity and speed of response the electrical disturbances were shown to be made up of small potential changes (about 50 mV) of short duration (about $\frac{1}{2}$ ms) and these electrical impulses were shown to travel along the nerves at speeds of up to 100 m/s.

The nerve impulse is the unit of nervous activity. Early in the present century Keith Lucas showed that nerve obeys the 'all or none law'. There is no relationship between the strength of the stimulus applied to the nerve and the magnitude of the action current; either the nerve responds to stimulation with a normal full-sized nerve impulse or it does not respond at all. This is true of all nerves, sensory or motor, whether excited directly, e.g. by electric shocks, or naturally by the sense organs.

Another important property of nerve, discovered in the early days, is refractory period. After an impulse has been evoked in a nerve it is not possible, however strong the stimulus, to evoke a second impulse within a period of about 1 ms. This is the absolute refractory period. It is followed by the relative refractory period, lasting 5–10 ms, during which the nerve recovers its excitability. These two properties, all-or-none response and refractory period, determine the way in which nerves can carry messages from one part of the body to another.

The elucidation of the nature of the nerve impulse is one of the recent triumphs of biophysics. It would have been impossible but for the discovery of the giant axons of cephalopod molluscs. In these animals the muscles of the mantle are innervated by axons which are several centimetres long and 500–1,000μ in diameter. At this size it is possible to put an electrode inside the axon, and it is also possible to extract the axoplasm in sufficient quantity for chemical analysis.

In outline, the story is as follows. At rest—that is, while no nerve impulses are passing—the surface membrane of the axon is continuously secreting sodium ion from the axoplasm to the blood, replacing it with potassium ion. As a result of this activity the concentration of sodium is lower (and that of potassium is higher) inside the axon than outside it; in addition, a difference of electrical potential is set up across the membrane, the inside being some 50 mV negative to the outside. The axon has something in common with an ordinary car battery in that it can be charged up and is then able to supply current; the process of charging is represented by the outward secretion of sodium and its replacement by potassium. Given the opportunity to do so—as when a hole is made in the membrane —sodium ions will diffuse inwards and potassium ions will diffuse outwards, the battery being thereby discharged.

By means of electrodes, one inside the axon and one outside, it is possible to pass current across some part of the membrane. The effect of a suddenly applied current directed outwards through the membrane (the inside electrode being made the anode and the outside electrode being the cathode) is to produce a local increase in the permeability of the membrane to ions, which amounts to the same thing as a local increase in electrical conductance. It is as though a hole has been made in the membrane, and this hole now provides a path through which ions can move so as to short-circuit the battery. The short-circuit current moves in through the active region (or hole), and to complete the circuit current must move out through the as yet unaffected neighbouring regions of the membrane (Fig. 24a). These outwardly directed currents then activate those parts of the membrane through which they pass, increase their conductance, and so on. In this way the active region (of increased conductance) sweeps over the membrane of the axon in all directions away from the point of stimulation, much as fire sweeps along a train of gunpowder. At any given place on the axon the active state lasts for about 1 ms. This is the absolute refractory period. As the conductance returns to its normal value the axon again becomes excitable.

The passage of a nerve impulse, then, is associated with a flow of current in local electrical circuits which are centred upon the boundary between the active and inactive regions of the axon membrane. The changes of electrical potential associated with these local currents can be detected and recorded on an oscillograph. Fig. 25 shows a record obtained in this way. The movements of ions which produce records of this type, and the changes in the electrical properties of the membrane which accompany them, have been worked out in some detail. What we still lack is an interpretation of these changing electrical properties in terms of the fine-structure of the membrane.

The velocity with which the nerve impulse travels depends among other things upon the diameter of the axon, the velocity being greater the greater the diameter. This is easily understood. The magnitude and extent of local action currents is limited by the resistance of the electrical paths, both inside and outside the membrane, and the electrical resistance of the inside path varies as the cross-sectional area of the axon. A few examples of the relation between velocity of conduction and diameter of axon are assembled in the accompanying table, which shows the expected correlation between velocity and diameter in non-myelinated axons. But in the case of myelinated axons the velocity of conduction is much greater than their diameter would suggest.

Nerve		Diameter (μ)	Conduction velocity (m/s)
Cat, saphenous	myelinated	14	80
sympathetic	non-myelinated	2	2
Frog, sciatic	myelinated	10	30
Crab, leg	non-myelinated	15	5
Loligo, giant axon	non-myelinated	700	22

This, too, is easily understood. The myelin sheath has insulating properties, and current can pass into and out of the axon only at the nodes of Ranvier. When the impulse has reached one node, and

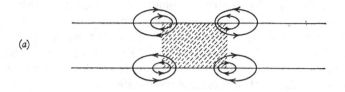

(a)

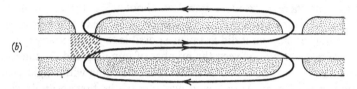

(b)

Fig. 24. Action currents in (a) non-myelinated, and (b) myelinated axons. The active region is shaded.

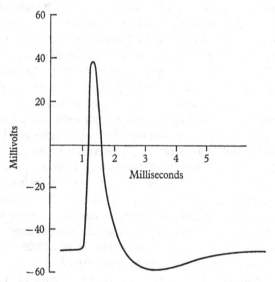

Fig. 25. Action potential recorded from an electrode placed inside the axon.

current is able to enter the axon at that point, the local circuit can only be completed if the outgoing current passes through the next node (Fig. 24b). In consequence, the impulse leaps from one node to the next instead of moving along the axon at uniform speed.

81

The knowledge that the nerve impulse is an all-or-none pheno-menon at once confronts us with another problem. We know that the contraction of muscles is not an all-or-none matter but that graded contraction is possible; and subjectively we know that our sensations are graded in intensity. If all nerve impulses are the same how is gradation achieved? How does the nervous system transmit differences in intensity? A likely way would be by variation in the number of nerve impulses transmitted and indeed almost all the early experi-ments in which sense organs were stimulated indicated in a general way that there were more impulses about when the stimulus was increased. But all this early work was done on nerve trunks contain-ing many axons and it did not prove possible to discover any quanti-tative relationship between the number of impulses and the intensity of stimulation. The technical difficulties of dealing with single axons were overcome by Adrian who was one of the first to apply valve amplification to physiological problems. There are in the striated muscles of vertebrates certain sense organs which are stimulated when the muscle is stretched. The sterno-cutaneous muscle of the frog is a very small muscle and contains only three or four sense organs. Adrian and Zottermann led off and amplified the impulses from the nerve supplying this muscle and then split the muscle down until only one sense organ and its axon remained intact. The impulses recorded from the nerve now showed a perfectly regular rhythm in contrast to the irregular discharges found in preparations with more than one sense organ in action. Graded stimuli were applied to the sense organ by stretching the muscle with different weights and it was found that as the weight was increased the interval between successive impulses in the nerve was decreased. The results of these experiments were published in 1926 and they establish the basic principle upon which the nervous system works—that variations in the intensity of a stimulus applied to a sense organ find expression as variations in the frequency with which nerve impulses pass up the sensory axon. This has been tested over a wide range of sense organs in vertebrates and in invertebrates and has been everywhere confirmed.

Now if on the sensory side increase in intensity of stimulation gives increase in frequency of nerve impulses, may we not expect that on the motor side increased frequency of impulses in the motor nerves will result in increased strength of contraction in the muscle? If we try to test this on nerve-muscle preparations from vertebrates we find that our expectations are not borne out. A single impulse in a motor axon evokes a twitch-like contraction of all the muscle fibres which it supplies. The motor unit in vertebrates is the single motor axon together with all the muscle fibres which it supplies. The graded contractions of the whole muscle are due to variations in the number of motor units which are active at any given moment. If frequency of nerve impulses really does determine the strength of the contraction then the conversion of frequency into number of active units must take place at an earlier stage in transmission, that is in the central nervous system.

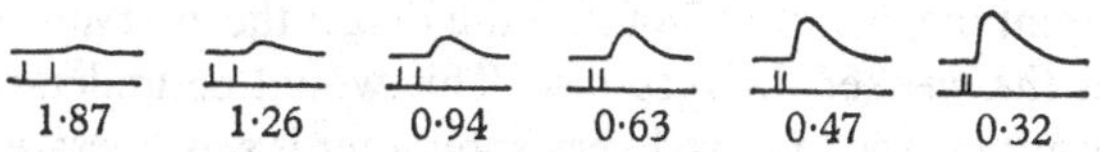

| 1·87 | 1·26 | 0·94 | 0·63 | 0·47 | 0·32 |

Fig. 26. Response of the sphincter muscle of a sea anemone to two electric shocks at different time-intervals apart.

But if we leave the vertebrates and turn to the invertebrates we find our expectations abundantly confirmed. In the invertebrates in general a single nerve impulse in a motor axon will not as a rule evoke a contraction in the muscle. At least two nerve impulses are required and the shorter the interval between them the more powerful the contraction. This is most easily seen in animals such as the sea anemones whose responses are very slow (Fig. 26). Furthermore, in the arthropods we very often find that all the muscle fibres in a muscle are supplied by the branches of a single motor axon. When we stimulate the axon electrically we find that a single shock fails to produce a contraction. The muscle will only respond to a train of impulses, and the tension developed by the muscle increases with the frequency of the shocks applied to the nerve. In these invertebrate muscles the arrival of a nerve impulse at the neuro-muscular junction, even though by itself it may fail to cause a contraction of the muscle, does nonetheless produce some effect which makes it possible for a second impulse to get through if it arrives sufficiently soon after the first. We call this effect facilitation.

The means whereby the arrival of a nerve impulse at the neuro-muscular junction stimulates the muscle fibre to contract has had a long history of controversy. Many muscles can be excited by electrical shocks applied to them directly, and according to one school of thought the nerve impulse acted as a straightforward electrical stimulus. The other school of thought held that when the impulse arrived at the nerve ending there was secretion of some substance which excited the muscle fibre. As far as vertebrate striated muscle goes this controversy has been settled, and we know what the chemical transmitter is; it is acetylcholine.

The concept of a chemical transmitter makes it possible to suggest an explanation of the phenomenon of facilitation. All nerve impulses being of the same size, let us suppose that each impulse produces the same quantity of chemical transmitter at the neuro-muscular junction. The quantity of transmitter produced by a single impulse is in most animals insufficient to stimulate the muscle fibre. But with the arrival of a second impulse the amount of transmitter produced by the two together may be sufficient. Our explanation must also take into account the observed fact that the longer the interval between the impulses the weaker the response. This would be understandable if the transmitter was removed very soon after it was liberated—and in the vertebrate body we know that acetylcholine is very quickly destroyed by the enzyme choline esterase. By analogy with other biochemical changes the simplest hypothesis is that the rate of breakdown of the transmitter is proportional to its concentration. If the nerve impulses arrive at a constant rate the concentration of the transmitter will be built up until its rate of breakdown is equal to the rate at which the nerve impulses renew it. In this way a greater frequency of nerve impulses will mean a greater concentration of transmitter.

It is unlikely that the sensitivity of all the muscle fibres to the transmitter will be exactly the same; it is far more probable that some will be more excitable than others. At a low frequency of stimulation the concentration of the transmitter will be low and only the most excitable muscle fibres will be stimulated. As the frequency is increased, the concentration will be increased and more muscle fibres will be stimulated. We call this recruitment. On this theory we can understand how frequency of nerve impulses in the motor nerve is converted into strength of contraction.

Where one nerve cell adjoins another, as for example where the axon of a sensory neurone makes contact with the dendrites of a motor neurone, there is a region of special properties known as the synapse. All the evidence indicates that conduction across synapses is of the same nature as conduction across the neuro-muscular junctions of invertebrates, at least in so far as the requirement for facilitation is concerned. This would lead us to suppose that facilitation is involved in any form of nervous transmission from one cell to another, be it from nerve to muscle or from nerve to nerve, and to suspect that the through-conduction of the vertebrate neuro-muscular junction is exceptional. Cases in which the muscle fibres respond to a single nerve impulse are also known in annelids, arthropods, and molluscs—but there they are the exceptions, not the rule. Such cases are in fact examples of the limiting case, the case in which the muscle fibres are sufficiently excitable to respond to the concentration of transmitter which a single impulse can produce. Lower its excitability with a mild dose of curare and the vertebrate muscle fibre will respond only to a train of impulses, no longer to a single one.

Another way of looking at the difference between vertebrate and invertebrate neuro-muscular systems is to see that in the invertebrate the muscle fibre is the motor unit, whereas in the vertebrate the motor neurone and the muscle fibres which it innervates are collectively the motor unit. In both vertebrate and invertebrate muscles gradation of contraction is brought about by changes in the recruitment of motor units; in one case this happens in the muscle itself, in the other case it must happen in the central nervous system.

Nerve impulses have now been detected in representatives of all the major phyla with the exception of the platyhelminthes, and failure here is almost certainly a matter of technical difficulty. Furthermore, nerves show the same essential properties of all-or-none response and refractory period in all animals in which they have been studied. There is therefore reason to suggest that, as in muscle, there is some fundamental activity common to all cells which has been specially developed in the nerve cells of animals. What might this be?

As we have seen, the passage of the impulse along the axon is associated with the passage of ions across the membrane of the axon. At rest, the axon expels sodium and takes in potassium to replace it; during the passage of an impulse these processes are momentarily reversed. In Chapter 4 we noted that all cells seem to maintain differences in ion concentrations across their membranes. All cells, then, have this much of the basic mechanism upon which the ability to conduct impulses depends. It may be that all cells have in some measure the ability to propagate changes in the permeability of their membranes; but we have too little information to make any firm statement about this.

8

SENSE ORGANS

In physiological experiments nerve fibres can be stimulated by almost any kind of rough treatment, by electric shocks, by application of certain chemical agents, by cutting or crushing. When functioning naturally in the body the sensory axons are activated by the sense organs to which they are connected. Like the nerves the sense organs can be stimulated by almost any sort of rough treatment, but it is the great virtue of the sense organ that it is very much more sensitive to one kind of treatment than to others. When we are able to measure accurately the strength of the stimulus, as we can do in the cases of sound and light, we find that the minimum amount of energy required to excite a single sense organ is of the order of 10^{-10} ergs or less.

Sense organs are classified in various ways. One system places them in categories according to the type of stimulation to which they are sensitive, as follows:

Sense organ	Sense served
1. chemoreceptor	smell and taste
2. mechanoreceptor	touch, pressure, tension; hearing and balance
3. photoreceptor	sight
4. thermoreceptor	sense of hot and cold
5. undifferentiated nerve ending	pain.

Alternatively they may be classified according to the location of the stimulus to which they respond. On this basis we recognize

(*a*) exteroceptors, supplying information about events at the surface of the body, e.g. the sense organs of touch, pressure, heat or cold, taste.

(*b*) proprioceptors, supplying information about the position and attitude of the body, e.g. the stretch-receptors in the muscles and joints and the vestibular apparatus of the ear which conveys the sense of orientation in the gravitational field.

(*c*) distance receptors; the sense organs of sight, smell and hearing. Although of course these sense organs are stimulated by events occurring at the surface of the body, e.g. by the impact of sound waves on the eardrum, we know subjectively that the mind associates the sound not with the place where it is detected but with the place where it originates.

(*d*) interoceptors; a relatively unimportant category of sense organs conveying sensations, mostly of pain, from the viscera.

There is no particular merit in these systems of classification and no advantage whatever in arguing about precise definitions. Such terms as are given above are in general use and are justified on grounds of convenience.

A more important property of sense organs which we will have to take account of is the property of adaptation. As described in the previous chapter Adrian's first investigations of single sense organs were carried out with the stretch-receptors in frog's muscle and he found that at a given load the corresponding frequency of nerve impulses was maintained more or less indefinitely.

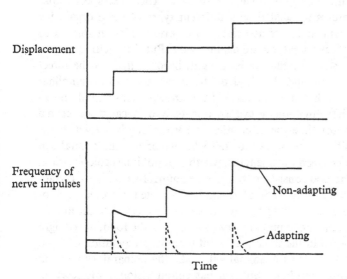

Fig. 27. Diagram to illustrate the difference between adapting and non-adapting sense organs.

When other sense organs such as those at the roots of hairs came to be investigated it was found that a displacement of the hair evoked a train of nerve impulses which rapidly decreased in frequency and soon ceased altogether. The sense organ at the base of the hair is sensitive to changes in displacement; it quickly becomes adapted to any constant displacement. The difference between adapting and non-adapting sense organs is illustrated in Fig. 27. In general, proprioceptive sense organs, which are largely concerned in postural mechanisms, are non-adapting—otherwise we could not stand up without conscious effort; exteroceptive sense organs are adapting—otherwise the central nervous system would be continually bombarded with trains of impulses from all parts of the body surface, most of which would be conveying useless information; what matters to the animal is not so much the precise form of the stimulation pattern it is receiving at any moment, but changes in the pattern which might call for action, and this requirement is neatly met by sensory adaptation.

There are three principal ways in which sense organs can be investigated. First there is the purely anatomical and histological method which, as far as it goes, is generally the easiest and quickest and often the only one which can be used. There are special anatomical features associated with different types of sense organ, for example, pigment layers associated with eyes, which enable us to decide that it is an eye we are dealing with. But this method cannot tell us how sensitive the eye is to light. Second, there is the direct physiological method. We lead off from the nerve to an amplifier and oscillograph and then allow light of known intensity and known wave-length to fall upon the eye. By this method we can discover in a relatively short time the intensities and wave-lengths to which the eye is sensitive. But we cannot tell whether or not the animal can distinguish between wave-lengths, whether or not it has colour vision. The third method consists in training the animal to associate a certain situation with a certain stimulus, e.g. to associate the presentation of food with a green light, so that upon seeing the green light the animal will make some move preparatory to accepting the food. A red light is then exposed in the same place, but without presentation of food, and generally at first the animal will make the preparatory move in response to the red light. But if it has colour vision it will learn to distinguish red from green. If so, we can then replace the red light with various shades of orange and so work towards green until the animal can no longer distinguish between the food signal and the no-food signal, and thus discover how small a difference in wave-length can be detected. This method gives all the answers; but it is extremely time-consuming and requires the most careful attention to the conditions of the experiment.

The words 'sense organ' are used both of the individual sensory element and of the complicated aggregations of sensory elements and ancillary apparatus such as the vertebrate eye. In the rest of this chapter we will not be concerned with sensory elements and the way they work; we will not inquire into the reactions by which a mechanical displacement appears in the nervous system as a train of impulses, but we will try to discover what sorts of mechanical displacement are likely to excite the mechanoreceptors in the ear, how these are related to the sound waves which the animal encounters, how far the ear is likely to be able to analyse complex sounds and so on. In fact, we will be concerned rather with the ancillary apparatus and with the performance of the sense organ as a whole. That being so, we will have nothing more to do with thermoreceptors and undifferentiated nerve endings and relatively little with the chemoreceptors, with which we now begin.

Organs of chemical sense

Our own chemical senses of taste and smell can be separated by reason of the fact that the sensations which we call taste come from substances in watery solution which are applied to sense organs in the mouth, whereas the sensations which we call smell come from substances which enter the nose as gases. We might also make the distinction that the olfactory epithelium of the nose is really part of the forebrain whereas the sensitive cells of the taste buds are developed from the general ectoderm like the cells of any other peripheral sense organ. But in the lower animals these distinctions cannot be maintained. In aquatic animals all chemical stimuli are waterborne. Senses of smell and taste are not necessarily confined to the head. Thus in some fishes taste buds are found all over the surface of the head and some way down the body, and in various insects there are organs of chemical sense in the feet. While we may find it convenient to retain the words taste and smell it is obvious that no precise significance attaches to them.

One of the difficulties attending the study of chemoreceptors is that we have no adequate means of controlling the stimulus and its application. For sound and light we have delicate instruments which can be used to check the intensity of the stimulus, but in the case of the sense of smell the sense organ is often literally millions of times more sensitive than any method of chemical analysis. The smell of skatol can be detected in concentrations of 0.4×10^{-6} mg/l. The well-known case of the male eggar moth which can detect the odour of the female at a distance of 2 miles must involve sensitivities of an even higher order. How is one even to begin to investigate such a problem with chemical techniques?

The range of substances which can be recognized is very great. In man it is reckoned at something between one and four thousand. Various attempts have been made to classify smells on a subjective basis, creating such categories as ethereal (fruit), aromatic (camphor), nauseating (carrion), etc., etc. What emerges from such classification is that there is no discoverable correlation between the chemical composition of a substance and the category to which it belongs. Very often chemical substances of widely different composition, such as nitrobenzene and benzaldehyde, are indistinguishable to the nose, whereas relatively minor variations in the molecule may remove a substance from one category to another. What does emerge, however, is that in bees, which alone among the lower animals have been intensively studied in this respect, the chemical substances tested fall roughly into the same categories as are recognized by man. This should encourage us in the belief that principles of general applicability can yet be established in this hitherto refractory field of investigation.

Of all the distance-receptive senses the chemical sense is the most primitive. A majority of animals rely upon it for finding their food, their mates and for avoiding their enemies. But of all the higher senses it is the least understood.

Organs of mechanical sense

Nerve endings differentiated in various ways are to be seen in the mammalian skin, and as a result of much detailed study we are now able to decide with fair certainty that certain types of nerve ending are normally excited by mechanical displacement and are associated with the sensations of touch or pressure. In most of the lower animals the sensory nerve endings show less differentiation and we are seldom able to decide that any particular type of nerve ending is susceptible to a particular type of stimulation.

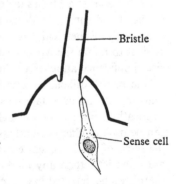

Fig. 28. Trichoid sensillum.

In the arthropods, and in particular in the insects, the problem is easier because if a sense cell is to be excited by mechanical displacement provision must be made for the transmission of such displacement through the otherwise rigid cuticle. The details of the sense organ's structure can often give us a good idea of how it works.

A very obvious case is that of the trichoid sensillum of an insect. This is an articulated bristle with a single sense cell of very characteristic form attached to its base (Fig. 28). As the bristle is moved in its socket it is clear that the sense cell will be subject to longitudinal push–pull displacements. If nerve impulses are recorded it is found that displacement of the bristle sets up trains of impulses in the nerve. The trichoid sensilla are like the sense organs of mammalian hairs in showing a rapid rate of adaptation.

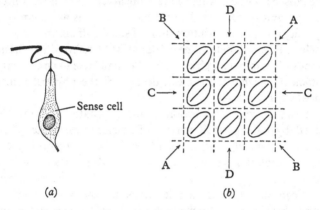

(a)   (b)

Fig. 29. Campaniform sensillum. (a) Single sensillum in section. (b) Diagram of group of sensilla; these will be stimulated by strains A and B, but not by strains C and D.

By contrast the fact that the campaniform sensillum is a mechano-receptor is not obvious and it was some time before this was discovered. The campaniform sensillum is seen as a small elliptical area of thin cuticle with perhaps a slight thickening or rib along its major axis. To the middle of this rib is attached a sense cell very similar to the one which is seen in the trichoid sensillum (Fig. 29a). Campaniform sensilla occur in groups, and all members of the group often have the same orientation. When closely packed they tend to be arranged as shown in Fig. 29b. Now as a result of this arrangement the mechanical properties of the cuticle are altered. Instead of being a sheet of uniform thickness having the same mechanical properties in all directions it is effectively transformed into a lattice. Such a lattice can be readily deformed by forces in the directions A and B but less readily by forces in the directions C and D. Deformation by a force in the direction A will cause the thin roof of the sensillum to bulge upwards and presumably a movement such as this will stimulate the sense cell. The campaniform sensilla are thus likely to be sensitive to certain directions of strain in the cuticle. All this has been abundantly verified by recording nerve impulses, and it has been shown that these are the proprioceptive organs of the insect which participate in its postural nervous mechanisms, to be described in the next chapter. In the campaniform sensilla, as we would expect in sense organs concerned in posture, adaptation is slow and incomplete.

The most elaborate organs of mechanical sense are those concerned in hearing and in balance. The mammalian ear is an organ of such complexity that any description short of some half dozen pages would be useless. For such details the reader must be referred to a standard text-book of mammalian physiology. We shall have to be content with a statement of what the ear can do and of the physical principles upon which it operates.

1. It can detect the direction of the gravitational field. This is achieved by having small masses of calcium carbonate (otoliths) attached to hair-like processes of sense cells. According to the animal's orientation the weight of the otolith pushes or pulls upon these processes.

2. It can detect linear accelerations by the same device. The otolith is heavier than the medium in which it is suspended and its inertia is greater. It will therefore tend to lag behind if the body is accelerated.

3. It can detect angular accelerations. If a vessel full of water is spun on its axis the water, owing to its inertia, does not immediately take up the motion of the vessel. This is what happens in the semicircular canals of the ear; the slip between the walls of the canal and their fluid contents deflects a small vane which is attached to sense cells.

4. It can detect sound waves and can carry out frequency analysis. The sound waves in the air are converted into mechanical vibrations of the eardrum and are transmitted by the ear ossicles to the cochlea. The operation of the cochlea makes use of the principle that high-frequency vibrations in a viscous fluid are more rapidly damped than low-frequency vibrations. Low-frequency vibrations travel further along the cochlea than high-frequency vibrations, and frequency analysis is thus made possible.

Of these four functions carried out by the ear the first three are concerned with balance, posture, and muscular coordination in general. In respect of these functions the ear is a proprioceptive organ, and these proprioceptive functions of the ear are well developed in all classes of vertebrates. Nearly all terrestrial vertebrates have an eardrum and either one or three ear ossicles, but the cochlea is hardly developed at all except in mammals and birds. Let us be clear that there is a vast difference between hearing and frequency analysis. The cochlea is essentially the organ of frequency analysis and while the lower vertebrates have no cochlea they can certainly hear. In fact any animal which has the means of detecting the gravitational field has also the means of detecting sound waves.

Better to understand that last sentence we will consider the workings of the simplest 'ears' in the animal kingdom. Sense organs, variously called statocysts or otocysts, are to be found in nearly all phyla even including the coelenterates. Fig. 30 is a diagram of a generalized statocyst. Its essential features are the statolith, a small particle of calcium carbonate, and the sense cells lining the vesicle, having hair-like processes upon which the statolith rests. The statolith, being of greater density than the fluid in the vesicle, comes to rest upon the hairs of the lower side and it is easy to see that this organ can appreciate the direction

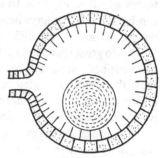

Fig. 30. Diagram of a simple statocyst.

of gravity in terms of the sense cells upon which the statolith rests at any given moment. As with the corresponding organ in mammals, acceleration in any direction will cause a movement of the statolith relative to the sense cells and this again will be capable of stimulating the latter. It also follows from simple physical laws, again owing to this same difference in density, that the passage of sound waves will cause relative movement between statolith and fluid and there-fore we must conclude that such an organ, if it is capable of re-sponding to gravity, must also be capable of responding to sound waves.

As far as is known no one has succeeded in recording nerve im-pulses in the nerve leading from a statocyst. But that such organs are responsible for orientation with respect to gravity has been demon-strated in the following ingenious way. In many crustacea such as the lobster there is a statocyst in the basal joint of the antennule. It is not a completely closed vesicle, but a pit open to the exterior, and it is lined with fine branched bristles to which sand grains are attached. When the animal moults the whole lining of this pit, including the bristles and the sand grains, is thrown off, and the animal is then faced with having to replace the sand grains. This it does by shovelling up sand with its claws and pouring the sand over its head. In this particular experiment the animal was allowed to moult in an aquarium in which iron filings took the place of sand, and when the moult was completed the animal duly shovelled the iron filings into its statocyst. An electro-magnet was then held over it and the animal at once began to swim upon its back.

Animal tissues have approximately the same density as water. Aquatic animals are therefore more or less permeable to the sound waves which reach them from the water, and their auditory organs can be sited almost anywhere within their bodies. But in the case of terrestrial animals the difference in density between the air and the body is so great that most of the energy of the sound waves is reflected from the body surface. The auditory organs of the terrestrial animal must therefore be located at its surface, or, if they are to be withdrawn from the surface, means must be found for allowing the sound waves to reach them. The fine bristles on the anal cerci of the cockroach are auditory organs exposed upon the surface. They are so light that they are capable of responding not only to gross movements of the air, but also to sound waves of up to 3,000 c/s.

Many insects have adopted the method used by terrestrial vertebrates, that is, the method of having a fine membrane, or tympanum, with air on both sides, stretched across the path of the sound waves. The locust has a pair of such organs in the first abdominal segment, the inner side of the membrane being open to the tracheal system. Applied to the inner side is a group of sense cells of that type which we find in other mechanoreceptors of insects (Fig. 31). By leading off impulses from the tympanal nerve it has been shown that this organ will respond to sound waves between 500 and 10,000 c/s. The advantage of this type of organ is that it can be withdrawn into the body with an air passage leading to the exterior and is thus protected from stimulation by contact and from damage.

In some insects—the order Diptera, including flies and mosquitoes —the second pair of wings is reduced to a pair of little clubs called halteres, which beat up and down synchronously with the wings when the insect flies. A fly deprived of its halteres becomes unstable in the horizontal (yawing) plane, which suggests that the halteres are organs of balance. As is well known, a gyroscope in rotation develops secondary forces in response to any displacement of its axis of rotation. The haltere may be looked upon as an alternating gyro, and according to theory it will develop secondary forces just like an ordinary gyro, except that the direction of the secondary forces is

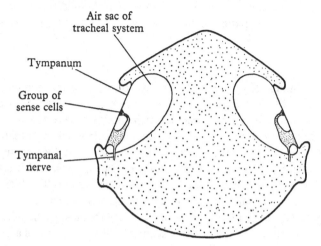

Air sac of
tracheal system

Tympanum

Group of
sense cells

Tympanal
nerve

Fig. 31. Transverse section through the first abdominal segment of a locust,
to show the tympanal organ.

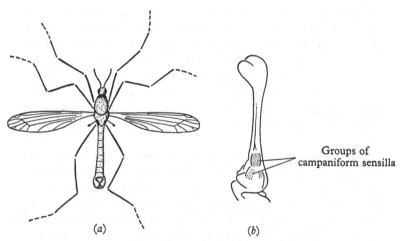

Groups of
campaniform sensilla

(a)   (b)

Fig. 32. Halteres. (a) Dorsal view of cranefly, to show halteres.
(b) Dorsal view of left haltere of a blowfly.

reversed upon the upstroke as compared with the downstroke.
Examination of the haltere shows groups of campaniform sensilla at
its base, so sited as to be stimulated by the secondary forces (Fig. 32).
To confirm this theory the haltere nerve was prepared for the recording of impulses and the preparation together with the leading-off

electrodes was mounted on trunnions so that it could be rotated; by such rotation the pattern of impulses was changed in the predicted manner.

Eyes

Many animals, not always primitive ones, have no recognizable eyes but yet have a diffuse sensitivity to light over the whole or over part of the body. Many lamellibranch molluscs have the habit of lying buried in sand or mud with only their siphons, or breathing tubes, projecting into the water. In the clam *Mya* a sudden change of light intensity results in the retraction of the siphons, and it seems that the photoreceptors concerned are single sense cells which are present in large numbers just under the epidermis. The earthworm has no eyes, but can be shown to be sensitive to light, particularly on the dorsal surface of the anterior segments. But such eyeless animals are usually burrowing in habit; in free-living animals of almost every phylum we find some kind of eye developed, even in the coelenterates. The eye is generally easily recognizable by the layer of pigment which screens the sense cells from incident light on all sides but one and so enables the animal to appreciate the direction of the light as well as its intensity. The pigment usually takes the form of a cup, varying in shape from the shallow saucer of the coelenterate (Fig. 33a) to the almost completely closed vesicle of the eye of *Nereis* (Fig. 33b). An eye of the latter type may well be compared with a pinhole camera. Unfortunately we have no information about the ability of animals possessing such eyes to perceive the form of objects.

Apart from the arthropods, whose compound eyes we shall consider presently, the eyes of most of the invertebrates are of about the same standard of complexity as the eye of *Nereis*. To this the cephalopod molluscs are an outstanding exception. Their eyes are on the vertebrate plane of organization and indeed show a remarkable superficial resemblance to the vertebrate eye, all the familiar features such as retina, cornea, lens, iris, and so on being clearly recognizable (Fig. 34). There is also a mechanism of accommodation, i.e. of adjusting the focus. The main points of difference between the cephalopod eye and the vertebrate eye can be traced to fundamental differences in development. One is that the cephalopod lens is formed out of two halves joined together, as can easily be seen in any section. A second point of difference is that the cephalopod retina is not inverted. The vertebrate retina is developed as an outgrowth of the

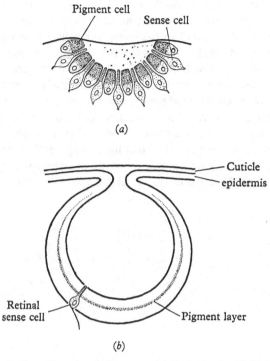

(a)

(b)

Fig. 33. Sections through primitive eyes. (a) Coelenterate. (b) *Nereis*.

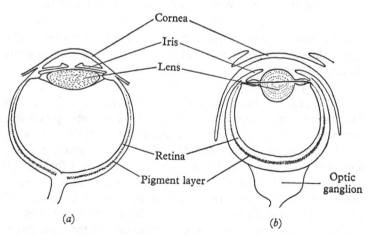

(a)   (b)

Fig. 34. Sections through well developed eyes.
(a) Vertebrate. (b) Cephalopod.

brain, the optic vesicle, which later assumes the form of a cup. The retina is formed from the inner wall of the cup and the nerve fibres run along the inner wall to the place where they pass out into the optic nerve. The rays of light have thus to pass through the nerve fibres before they reach the rods and cones which are the photo-receptors in the retina; for this reason we speak of the retina as being inverted. In the cephalopod eye the retina is developed as an invagination from the ectoderm of the surface of the head and the nerve fibres leave the retina on its outer (convex) side so that the photoreceptors are directly exposed to the incident light—apparently a much neater and more logical arrangement. From its structure, the cephalopod eye should be the equal of the vertebrate eye. The final resemblance

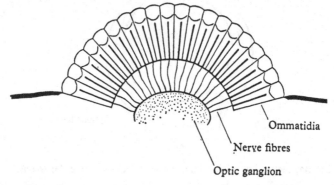

Ommatidia

Nerve fibres

Optic ganglion

Fig. 35. Section through the compound eye of an arthropod.

between these two types of eye, notwithstanding their different origins, makes this one of the most striking cases of convergence in evolutionary history.

The compound eyes of arthropods work upon an entirely different principle. Instead of having a retina in the form of a cup and a lens focusing an image upon it, the compound eye is made up of a large number of small elements known as ommatidia. A single ommatidium is shown in Fig. 36. Over the distal end the cuticle forms a clear cornea and under this is the crystalline cone, or lens, which focuses light upon the end of the rhabdome. This rhabdome is a rod-like structure formed by contributions from each of seven or eight retinal cells which surround it. How far each retinal cell and its rhabdomere can respond independently of the other retinal cells in the same ommatidium is not known with certainty; for the present we shall make the simplest assumption, namely, that the retinal cells of a single ommatidium act as a unit. The ommatidium is separated from its neighbours by two rings of pigment cells, proximal and distal. Each ommatidium may be looked upon as a light-tight tube having a light-sensitive element at the bottom, which can only be reached by light entering more or less parallel with the long axis of the ommatidium.

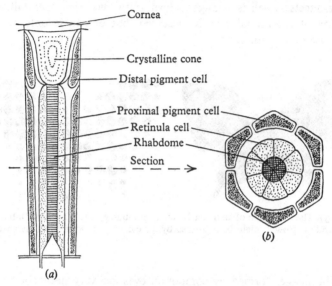

Fig. 36. Ommatidium from the compound eye. (*a*) Longitudinal section. (*b*) Transverse section.

In the simplest compound eyes such as that shown in Fig. 35 there is a small number of ommatidia and the angle subtended by each is relatively large. An eye like this will be able to appreciate the environmental situation in terms of light in some directions and darkness in others, but it does not seem as if it will be capable of forming anything which would be worthy of the name of image. In the more advanced compound eyes, such as those of dragonflies, the number of ommatidia is increased (up to 28,000) and the angle subtended by each ommatidium is reduced down to 30′). Knowing this angle we can reconstruct the sort of image which such an eye as this might form. The image of course will be made up of a pattern of spots of varying brightness, one spot for each ommatidium (Fig. 37).

Under conditions of low light intensity the amount of light incident upon a single ommatidium of a compound eye may be insufficient to stimulate the retinal cells. It is commonly found that compound eyes undergo a process of dark-adaptation, which consists in the movement of the pigment within the pigment cells towards the two ends of the ommatidium. The ommatidium when dark-adapted is no longer a light-tight tube, and rays of light can reach the retinal cells through other facets as well as through its own. In this way the general illumination of the retinal cells is increased, but of course only at the expense of definition.

(a)   (b)

Fig. 37. The formation of an image by the compound eye. (a) Outline of a butterfly, and (b) How it might be registered by the eye of a dragonfly at a distance of 10 in.

The images formed by compound eyes are very poor compared with those which our own eyes can form. In the last resort visual acuity and the quality of the image must be limited by the angle subtended by the photoreceptor unit. In the human eye the angular subtension of a single cone is less than 1', whereas in the dragonfly, as mentioned, the angle subtended by a single ommatidium is about 30'.

This short survey has covered no more than a few selected examples from the very great range of sense organs found in the invertebrates, but for our present purpose there would be very little advantage in extending it. As with the contraction of muscle or the conduction of impulses by nerve, so also in the stimulation of photoreceptors by light it may well be that we are dealing with biophysical and biochemical processes which are common to all animals. But there is no doubt that in building up large and complicated sense organs to make best use of the photoreceptor animals do not seem to be bound down by any rules. On the one hand we can point to the remarkable convergence between the cephalopod eye and the vertebrate eye, on the other hand we can point to the semicircular canal and the haltere as organs playing much the same part in dynamic balance yet making use of completely different mechanical principles.

What is perhaps of more immediate importance is to consider the range of sensitivity of the lower animals' sense organs in relation to that of our own. The range of the human ear is from 16 c/s to 20,000 c/s in youth, but for other animals the audible range may lie between different limits. The bat can produce and hear sounds of about 80,000 c/s and makes use of them for locating obstacles in darkness on the 'asdic' or 'sonar' echo-ranging principles. A fish on the other hand is likely to be interested in sounds of about 2 c/s such as might be caused by worms wriggling in the water. In trying to understand the part played by hearing in the behaviour of animals one must always bear in mind that they may be responding to sounds which are quite inaudible to us. In some cases at least we can be fairly certain of what the lower animals' sense organs cannot do. It is extremely improbable, for example, that any other animals can carry out frequency analysis in the way that birds and mammals can. But this does not mean that they cannot use sound as a means of communication. Our ears are particularly sensitive to change in the pitch of a note whereas in insects such as the locust the ear has been shown to be sensitive to changes in loudness. The females of certain grass-hoppers are attracted to the chirps of males transmitted by telephone and suffering thereby such distortion as to be unrecognizable to the human ear. It seems that in these animals the rhythm of the sound rather than its pitch is the significant quality in giving meaning to it.

So also with light. The range of the spectrum visible to the human eye is from 4,000 Å to 8,000 Å approximately, with maximum sensitivity about 5,500 Å. The corresponding range for the bee is 3,000 Å to 6,500 Å. This is worth knowing because it means that in red light we can observe the behaviour of bees inside a hive under conditions which correspond to complete darkness from the bee's point of view. On the other hand we have to remember that objects which look to us to be of the same colour may have very different powers of reflecting ultra-violet light and so may appear very different to the bee's eye. Colour vision is of quite sporadic occurrence in animals. In the vertebrates it is found in the bony fishes, in lizards, in birds, and in man and the monkeys among mammals. It is not found in the dogfish, nor in the frog, nor in mammals other than the primates. Among the invertebrates colour vision has been demonstrated in bees and in some crustacea. But the only way to discover whether an animal has colour vision or not is by means of very laborious training experiments and up to the present not many invertebrates have been so tested.

In the next two chapters we shall have to give further consideration to the sense organs in their relation to coordination and behaviour.

9

NERVOUS COORDINATION

The philosophers of the Ancient World and the Middle Ages were in general agreed that the nervous system was concerned in perception and in the coordination of movement, and that the brain was the seat of the higher faculties. This, however, fairly represents the limit of their progress towards useful knowledge. Their attention was mainly occupied in fruitless speculation about the localization of function in the brain and about the seat of the soul in particular. In any case, in their day, investigation of this subject could hardly expect to prosper since the investigators could take account only of the shape of the brain whereas the really significant anatomical features of the central nervous system (C.N.S.) are only to be discovered with the aid of the microscope. The nervous system is difficult material for histological study and it was not until the second half of the nineteenth century that the tracts of nerve fibres running through the C.N.S. could be traced from their origins to their destinations. As this anatomical knowledge accumulated it became possible to build up a conception of central nervous function in terms of the paths up which sensory impressions would spread or down which motor control was exercised; but analysis of events in the C.N.S. made little progress. The concept of the reflex belongs to the first half of the nineteenth century, but it was not until its close that Sherrington's work on reflexes got into its stride. Sherrington investigated the reflex responses of spinal dogs, that is, dogs in which the spinal cord was transected in order to isolate the trunk and limbs from the influence of the brain. His experimental technique was very simple; the merit of his experiments lay in their design and interpretation. His account of them, published in 1906 as *The Integrative Action of the Nervous System*, pointed the ways which investigations of the C.N.S. were to follow in the next two decades. It is interesting to reflect that this line of investigation was successfully pursued before any clear ideas had been reached about the process of conduction in nerve.

Sherrington's experiments will serve very well as our introduction to central nervous function. We will, however, take advantage of modern knowledge in interpreting them.

If we apply a stimulus which is calculated to be painful to the foot of a spinal dog, the foot is pulled away. This response is a reflex, the simplest act of integration which the C.N.S. can perform, and very obviously an act of adaptation by the body to a change in its environment. We are certainly at liberty to suppose that when the stimulus is given trains of nerve impulses run up the sensory nerves to the C.N.S. and are there applied to the dendrites of the motor neurones which innervate the flexor muscles of the leg. Some of these motor neurones are excited, impulses are set up in the motor axons, the muscles contract and the foot is withdrawn. The nervous path involved in a reflex is called a reflex arc.

If we were to cut the sensory nerve between the sense organ and the C.N.S. and stimulate the central end with electric shocks, the withdrawal of the foot would follow as before. But if we cut the motor nerves and stimulated the central ends of these there would be no response in any part of the animal, and if we went further and placed electrodes upon the sensory nerve we would not be able to detect any impulses in it. The reflex arc works only one way, that is to say it has the property of polarity.

You will remember that Adrian used for his experiments a sense organ which was located in the substance of a muscle and was sensitive to stretch. Such stretch receptors are present in all vertebrate skeletal muscles. If we raise the foot of a spinal dog so as to flex the leg we find that the leg muscles respond with an extensor thrust. What happens is that in flexing the leg we stretch the extensor muscles and stimulate the stretch receptors. Trains of impulses pass up the sensory nerves to the C.N.S., pass over to the motor neurones which innervate the corresponding muscles and produce contraction of the muscles in the usual way. The knee-jerk, used in medical practice as a test of central nervous function, is this very same reflex. When the patellar tendon is struck the extensor muscles of the knee suffer a slight tug; that is enough to stimulate the stretch receptors and to evoke the contraction of the muscles.

Now let us see how this particular reflex plays its part in coordinated movement. Returning to the spinal dog, if we allow the posterior part of the body to rest upon the hind legs we shall observe that the legs, originally limp, yield a little as the weight comes upon them, then stiffen and support the body. This is the result of activity in the reflex arc we have been discussing; as the legs yield the extensor muscles are stretched and the reflex contraction is evoked. If a moderate weight is placed upon the dog's back the legs will yield a little further and the contraction of the extensor muscles will be reinforced. Reflex arcs of this sort are fundamental to the maintenance of posture in the animal. We may imagine that for as long as the animal remains standing there will be trains of impulses passing continually up the sensory nerves and down the motor nerves, the frequency of the impulses varying according to the weight carried by the leg and the corresponding state of contraction in the extensor muscles. Reflexes, then, are not to be thought of as events of limited duration; they can, and often do, involve sustained activity for long periods. Sherrington called this postural mechanism 'the spinal arc of plastic tonus', meaning that it served to maintain a degree of contraction, or tone, in the muscle which could be varied according to the circumstances of the moment. Although there is no fundamental difference between tonic reflexes which are continually at work in the maintenance of posture, and phasic reflexes which are events, we shall find this distinction helpful in appreciating the ways in which central nervous activity is adapted to the situations which arise in the normal life of the animal.

If while the leg is thus supporting the weight of the body we apply a painful stimulus to the foot, the leg is flexed at once. The withdrawal, or flexion, reflex in response to the painful stimulus completely overrides, or inhibits, the tonic extension reflex. As soon as the painful stimulus is removed the extension reflex returns. Now this property of inhibition is of the very greatest significance in the coordination of movement. When two stimuli evoking incompatible responses are applied simultaneously the response appropriate to one appears in full vigour while the response to the other is completely suppressed.

As yet it is not possible to give a convincing account of inhibition in the vertebrate nervous system. In vertebrates inhibition is found only inside the C.N.S., and what happens inside the C.N.S. is very much more difficult to investigate than what happens peripherally. But in many invertebrates inhibition takes place at neuro-muscular junctions and is accessible for study; some inhibitory chemical transmitters have been identified.

Finally, there is further information to be gathered from the experiment referred to in the last paragraph but one. When the painful stimulus is applied to one foot and this is withdrawn the animal does not necessarily collapse. A stronger extension is set up in the other hind leg. This of course is what we might expect, for when the whole weight falls on one leg it will yield a little and increase activity in the reflex arc. But what is more significant is that if we repeat the experiment with the animal supported so that its legs do not touch the ground, the flexion of one hind leg is accompanied by an extension of the other, such as would anticipate the extra load to be carried. And if we look further we can see that appropriate adjustments are made in the fore legs as well. Hitherto we have talked of simple reflexes and of simple reflex arcs as if they were events and paths which were independent of other events and unconnected with other paths in the C.N.S. This idea must be given up. The simple reflex is a convenient but artificial abstraction. In such an animal as the dog there are very few movements made by one leg which do not involve compensatory movements in the others. And while we may retain the conception of the single muscle and its spinal arc of plastic tonus we must look upon such reflex arcs as being coupled up with the corresponding reflex arcs of other muscles and, by virtue of these couplings, of being affected by nervous events in distant parts of the body.

Even if we had complete knowledge of how nerve impulses are transmitted from cell to cell in the C.N.S. we would not be very much better off in trying to understand the coordination of movements, even of simple movements. We should do well to realize that in view of the extensive interrelations which now begin to reveal themselves, any interpretation of central nervous function in terms of nerve impulses, facilitation, recruitment, etc. would be too complex for the mind to grasp. For the moment let us be content with general impressions and see whether a study of the lower animals will give us greater insight into these problems.

The type of coordination in which we are interested is that whereby the nervous system brings the effector organs—the muscles—into appropriate functional relationship with the sense organs. In some of the most primitive animals this is done by means of a nerve net. The nerve net of *Hydra*, a familiar illustration in elementary textbooks of zoology, is shown connecting up sense organs and contractile muscle tails more or less at random. Yet it must be capable of some discrimination if it is able to coordinate movements which can lead to the capture and ingestion of a small crustacean. But clearly this type of nervous organization has very limited potentialities. The analogy of the telephone exchange at once springs to mind. If the nerve net arrangement were adopted in the telephone system it would be impossible to call a distant subscriber without disturbing the peace of others living on the same route; this is of course avoided by having a central exchange. The significance of centralization is further seen by analogy with an army. The problems of coordinating the muscles of the body are very much the same as those of coordinating the activities of any large organization such as an army and involve the same hierarchical system, with local headquarters subordinated to higher formations and so on. This is only possible if there is centralization of authority and recognized channels of communication.

The nerve fibres of the coelenterate nerve net are thin and extremely difficult to see in the living animal, and the technical difficulties of placing micro-electrodes upon them are formidable. Nevertheless it has proved possible to record nerve impulses from the coelenterate nerve net. But what we know about the working of the nerve net comes from direct observation and mechanical recording of the responses to natural stimuli and to controlled electric shocks. But let us first try to predict, on the basis of our knowledge gained from the higher animals, what is likely to happen in the nerve net and then compare our prediction with the facts.

Let us suppose that the cells of the nerve net make synaptic connexions with one another and that conduction across the synapses and neuro-muscular junctions involves facilitation. Let us consider a sense organ in direct functional connexion with one nerve cell of the net. The sense organ is stimulated and generates a train of impulses. The first impulse will travel as far as the synapse between the sense organ and the first nerve cell, but no further. The second impulse, taking advantage of the facilitatory effect of the first, will pass to the nerve cell and will travel all over it, as far as the synapses which separate that nerve cell from its neighbours in the net. The third impulse will pass over to each of the neighbours, and so on.

Now let us get a sea-anemone and discover what really happens. We stimulate a tentacle by scratching it lightly with a needle. The tentacle shortens slightly and bends towards the mouth. We continue to stimulate. Neighbouring tentacles join in the same movement. Soon the edge of the disc at the affected place is raised up and finally the edge of the disc rises up all round, the centre of the disc sinks down and the anemone closes up. Not in detail, but in general terms things have turned out as we expected; we have seen a zone of muscular activity slowly spreading from the point of application of the stimulus. This type of response is called decremental conduction.

This is not the only type of response which the sea-anemone can exhibit. If instead of stimulating a tentacle with a needle we take a blunt glass rod and press down firmly upon the edge of the pedal disc we will presently see the animal close up rapidly and completely. In this movement there is no suggestion of a gradual spread of the contraction away from the point of stimulation—the response is synchronous and symmetrical about the axis of the body. It is obvious that this response could not be coordinated by the sort of nervous mechanism which we originally supposed the sea-anemone to possess. We have to suppose that within the nerve net, or in addition to it, there must be tracts along which nerve impulses can be more rapidly

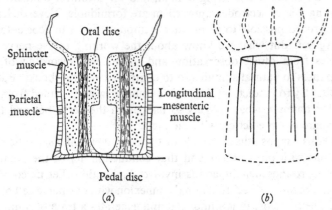

Fig. 38. Neuro-muscular system of the sea-anemone. (a) Vertical section, to show the principal muscles. (b) Diagram to show the disposition of the through-conducting system.

conducted. These tracts can be demonstrated most easily by physiological methods. It is possible to measure the conduction time in various directions and such measurements show that the velocity of conduction around the surface of the column is 10–20 cm/s, while the velocity of conduction in the mesenteries and round the oral disc is more than 1 m/s. Such physiological investigation indicates that the through-conducting system is arranged as in Fig. 38b.

There is histological evidence that in the mesenteries the fibres of the nerve net are elongated in the direction of the through-conducting tracts. But it does not seem likely that any single nerve fibre extends from one end of the body to the other. It is more probable that in the through-conducting system we have to do with a part of the nerve net in which the synapses are permanently facilitated, in the same way as the neuro-muscular junctions of vertebrates. But although the synapses may be permanently facilitated, the characteristic signs of facilitation and its dependence upon frequency appear at some point between the through-conducting system and the contractile elements which it serves. This can be shown by electrical stimulation of the through-conducting system at different frequencies.

The muscular system of the sea-anemone is illustrated in Fig. 38 a. Although coelenterates are supposed to be two-layered animals and therefore ought not to have muscles in the gross anatomical sense of the word, it is open to them, without infringing the principles of morphology, to thicken up their tissues by folding the layers which are nominally one cell thick. The principal muscles thus formed are (1) the parietal muscles running in the mesenteries close to the column wall, by whose contraction the column is shortened; (2) the longitudinal mesenteric muscles, running in the mesenteries between oral disc and pedal disc, by whose contraction the oral disc and tentacles are pulled down; and (3) the sphincter muscle running around the top of the column, by whose contraction the column is closed over above the oral disc. In addition circular muscle fibres are more or less uniformly distributed in the wall of the column.

If we apply electrodes to the wall of the column and through them deliver shocks at a frequency of 1 in 10 s there is no visible response for some time. Then gradually a constriction appears around the wall under the electrodes; this is due to the contraction of the circular muscle. At a frequency of 1 in 5 s the parietal muscles begin to contract slowly and the column shortens; but the oral disc and tentacles remain exposed. When the frequency is increased to 1 in 2 s there is a distinct contraction of the longitudinal mesenteric muscles, and the oral disc and tentacles begin to sink down into the column, but the sphincter remains inactive. Finally if the frequency is increased to one shock per second the oral disc and tentacles are pulled rapidly down and by contraction of the sphincter the delicate structures are covered over by the tough column wall.

There are two points to be noted about this experiment and its results. First, facilitation is involved in the excitation of all these muscles; in no case is a single shock effective. Second, the different muscles have different excitabilities (as determined by the minimal frequencies required to excite them) and the differences in excitability from one type of muscle to another are clearly of importance in the mechanism of coordination. It would be useless for the sphincter to contract before the longitudinal mesenteries had pulled down the oral disc and tentacles; their different facilitation rates ensure that the movements do not take place in the wrong order. In such ways as this coordinated and adaptive movements can be brought about by taking advantage of differences in facilitation rate.

In order to complete the coelenterate picture, let us add briefly that in the jellyfishes the development of the through-conducting system reaches its natural limit in that the neuro-muscular junctions as well as the synapses are permanently facilitated and a single shock applied to the nerve net will produce a contraction in the whole musculature.

As remarked earlier, the coelenterates are difficult animals from the physiologist's point of view, and the technical difficulties which their study presents make it unlikely that anything will come out of the coelenterates to affect the investigations of nervous phenomena in general. Rather it is the other way round. Our ideas about the nervous system and the way in which it acts have been derived from studies of the vertebrates and have proved themselves useful in interpreting the responses of coelenterates. That in itself, and in its support for the idea of fundamental uniformity in the nervous mechanisms of animals, is a matter for satisfaction. But we may note that many of the properties of the mammalian spinal cord have not been discovered in the sea-anemone. To begin with, the nerve net is not centralized in the anatomical sense; polarity is very weakly developed, inhibition not at all. We ought to hesitate, however, before coming to the conclusion that the sea-anemone represents a stage in evolution before the property of inhibition appeared. Inhibition is called for when muscles are so disposed as to act in antagonism. If we look into the arrangement of the muscles we shall see that in these relatively violent responses which have concerned us the contraction of one set of muscles does not automatically entail the relaxation of another set. When the sea-anemone shuts down it lets the fluid escape from its interior; when it expands again it pumps itself up by water currents driven by cilia. *Hydra* on the other hand does not take such liberties with its hydrostatic skeleton. The longitudinal and circular muscles are clearly in antagonism here and it would be interesting to know if reciprocal inhibition of the antagonistic muscles is developed in *Hydra*.

The next level of organization which we will consider, but only briefly, is that of the echinoderms. In all the echinoderms a sub-epidermal nerve net is very much in evidence. It is best seen at work in the sea-urchin. If a point on the surface of the sea-urchin is stimulated with a needle the spines and pedicellariae (little pincer-like organs) bend towards the source of stimulation. The longer the stimulus is continued, the further does the response spread away from the source—typical decremental conduction.

We also see in the echinoderms the beginnings of nervous centralization in the anatomical sense and the appearance of some of the physiological properties which we associate with the C.N.S. In a starfish the C.N.S. takes the form of nerve strands running down the groove in the under surface of each arm and forming a ring around the mouth. These strands are little more than condensations of the nerve net, but in close association with them are the motor nerve cells which supply the muscles. The organization of the system, as central nervous systems go, is pretty low; nevertheless the coordination of the animal's movements can be shown to depend upon its integrity. The starfish walks by means of its tube feet—there are hundreds of them—which are found on the under sides of the arms. If the starfish is turned upon its back and so supported that its tube feet do not come into contact with any solid object, the tube feet perform stepping movements. One arm—it may be any one of the five—is temporarily in the lead; its tube feet step parallel with its long axis. In the other arms the direction of stepping is parallel with the long axis of the leading arm and may make any angle with the axes of the other arms. If the leading arm is given a violent stimulus, stepping ceases and then presently starts up again with another arm in the lead and all the tube feet stepping parallel as before, but in the new direction. If a single cut is made in the circumoral ring of the nervous system the coordination of the tube feet is unimpaired, but if two cuts are made so as to isolate two arms from the remaining three, the tube feet within the two-arm portion step parallel and those within the three-arm portion step parallel, but the direction of stepping is not necessarily the same in the two portions.

The ability of one arm to take the lead and to dominate the others implies that the property of inhibition is present in the starfish nervous system. In this respect it does better than the sea-anemone. Polarity in the nerve net is better developed. But the study of the echinoderm C.N.S. is still very much at the anatomical stage. Nerve impulses have been detected in echinoderms but the effects of electrical stimulation have proved difficult to interpret.

The nervous systems of coelenterates and echinoderms are too low in organization and too difficult to study for us to have much hope of finding clues as to the way in which the central nervous properties emerge from the simple association of sense organs, nerves, and muscles. In any case these are radially symmetrical animals and may very well have different problems and different ways of going about them, so that perhaps there is less useful information to be gathered here than at first we thought. In the platyhelminths the technical difficulties are at least as formidable as in the coelenterates and echinoderms and at present they have very little to contribute. It is therefore to the annelids that we will turn next.

The annelid C.N.S. consists of the brain and the two ventral nerve cords, generally bound together, with ganglia in each segment. It is not buried in a network of muscle fibre and parenchyma, as is the platyhelminth C.N.S., but lies exposed in the coelom. The nerves proceeding from it to the body wall are short, but just long enough for leading off action currents. Among annelids the earthworm is a good experimental animal and a good one for us to begin upon.

The mode of progression of the earthworm has been described as peristaltic. Certainly it involves waves of contraction passing along the body which give it the appearance of a length of intestine undergoing peristalsis, but it is just a little bit more complicated than intestinal peristalsis, especially in that it has to be combined with some method of getting a grip on the substratum. We will look into this a little further.

The body cavity of the earthworm is divided up transversely by the septa. Each septum is a sheet of muscle which completely separates one segment from the next except for a small hole on the ventral side. This hole is surrounded by a sphincter-like concentration of muscle which is normally fully contracted so that the septa are to all intents and purposes complete. If the septa were rigid watertight bulkheads the pressure set up in the coelom of any segment would be determined only by that segment's own musculature. Pressure set up by the muscles in one segment could not be transmitted along the body and used to stretch muscles in another part. The septa are watertight bulkheads but of course they are not rigid. Nevertheless their presence affords the possibility of local differences in pressure which could not occur if the fluid in the coelom were completely free to move up and down the body. The presence of the septa affords some degree of independence to the individual segments.

Fig. 39 is a diagram showing successive stages in the progress of an earthworm as seen from the side. In position a segments 2, 3, 4, 5, and 12, 13, 14, 15 are in contact with the ground. Their circular muscles are relaxed (longitudinals contracted) and the chaetae are protruded. Segments 6, 7, 8, 9, 10, 11, and 16, 17, 18, 19, 20 are shown out of contact with the ground, their circular muscles are contracted and the chaetae are withdrawn. These regions are moving forwards; the longitudinal muscles of 11 and 12 are in process of relaxing and those of 7 and 8 are in process of contracting. Segments are thus being continually added to the posterior sides of the regions of contact and taken away from the anterior sides. As a result of this the regions of contact move slowly backwards relative to the ground, but the body of the worm moves more rapidly forwards relative to the regions of contact, so that there is a net forward movement.

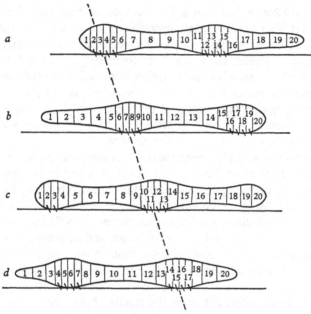

Fig. 39. Successive stages in the crawling of an earthworm.

Now let us try to invent a nervous mechanism for the earthworm as we did for the anemone and see how it compares with the real thing. Let us suppose that the earthworm has stretch receptors in its muscles and that these are connected with the motor neurones of the same muscles in a series of reflex arcs of plastic tonus. Further let us suppose that the reflex arcs of the circular muscles are coupled with the reflex arcs of the longitudinal muscles (in the negative sense that activity in one inhibits activity in the other) and that there is no nervous connexion between one segment and the next. (We know this isn't true, but let it pass for the moment.) A stretch applied to the longitudinal muscles of one segment will evoke a reflex contraction in the same muscles. The contraction of the muscles of one segment will stretch the muscles of the next segment, they will contract in their turn and so on. In this way a wave of contraction will be propagated along the body of the worm, coordinated by no more elaborate a nervous mechanism than the one we have just postulated. Nor is such a mechanism purely hypothetical. There is independent evidence that stretch receptors do exist in the body wall, because impulses can be detected in the nerves when the body is stretched. It is also possible to show that if the whole longitudinal musculature of the worm is stretched a reflex contraction is evoked. Finally there is a very old but none the less striking experiment, first performed by Friedlander in 1888. The worm is cut completely in two and the two halves are sewn together; the waves of contraction are then seen to pass over the cut as if it were not there at all. It can therefore be argued fairly that intrasegmental reflex arcs, acting in conjunction with the purely mechanical transmission of forces, will account for the peristaltic movements of the earthworm.

Now we know that intersegmental nervous connexions do exist and may well imagine that they are not without some part in the peristaltic movements, the foregoing demonstration notwithstanding. In 1904 Biedermann made an experiment which is more or less the exact opposite of Friedlander's. He cut the worm in half except for the C.N.S. which was the only tissue left intact, and he pinned down the two halves on either side of the cut so that the one could not pull on the other. Once again the waves of contraction passed freely over the cut.

There is no contradiction in the results of the two experiments. Both mechanisms can and do exist. This is simply an example of double assurance. Alternative paths are the rule rather than the exception in the C.N.S. and we find in general that if we sever the primary paths the animal is not put out of business completely; it turns to other devices using different channels.

At this stage we might usefully turn our attention to the earthworm's C.N.S. and to the connexions of the nerve fibres which are associated with it. Apart from the giant fibres, of which there will be more said in a moment, the nerve fibres of earthworms are short in relation to the body. The cell bodies of the motor neurones are grouped together in the segmental ganglia, their axons supplying the muscle fibres of the same segment. The cell bodies of the sensory neurones lie peripherally, near the sense organs, and their axons run into the C.N.S., generally terminating in the ganglion of the same segment. In addition there are many internuncial neurones, neurones whose axons are wholly confined to the C.N.S., running a short distance along the cord from one ganglion to another. These nervous connexions, illustrated in Fig. 40, form an adequate anatomical basis for the interpretation of Biedermann's experiment; we can imagine that stimulation of the stretch receptors of one segment excites not only the motor neurones of the same segment but also, via the internuncial neurones and with consequent synaptic delay, the motor neurones of the next segment, and in this way the wave of contraction passes smoothly over the cut.

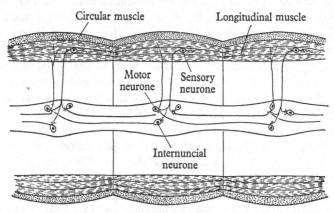

Fig. 40. Diagram of nerve connexions in an earthworm.

The rate at which excitation is transmitted from neurone to neurone along the nerve cord of the earthworm is about 50 cm/s, not very much faster than decremental conduction through the nerve net of a sea-anemone. This is adequate for the coordination of the peristaltic waves, but as in the case of the anemone there is a requirement for a through-conducting system to look after escape reactions, some system that will get through to the appropriate muscles in the shortest possible time. In the earthworm the through-conducting system is easily recognized. It consists of three giant fibres running the length of the body; they lie on the dorsal side of the nerve cord and can be seen in any section. The median fibre is of greater diameter than the two laterals, about 50μ as compared with about 25μ, and conducts more rapidly, about 30 m/s as compared with about 10 m/s.

These giant fibres are of the syncytial type, that is to say they are formed by the fusion of the axons of many nerve cells distributed along the cord. Nerve fibres run from the giants to the longitudinal muscles of each segment. The protoplasm of a giant fibre is not continuous throughout its length: there are partitions or giant synapses between the segments, but these partitions do not hold up the nerve impulses, which are conducted freely from end to end. The two lateral giants are connected together and an impulse set up in one spreads to the other, but the median giant is physiologically isolated. There are thus, in effect, two separate through-conducting systems; and since the neuro-muscular junctions between the giant fibres and the muscles are of the permanently facilitated (vertebrate) type, any impulse set up in any part of either system will produce a contraction of all the longitudinal muscles of the body. What then, we may ask, is the point of having two such systems? The point is that although the two systems have the same connexions with the longitudinal muscles they have different connexions with other parts of the body. It can be shown that if a stimulus is applied to the sense organs of the head end impulses are set up in the median giant, whereas if a stimulus is applied to the tail end impulses are set up in the lateral giants. Further, stimulation of the median giant causes protrusion of the chaetae at the tail end and stimulation of the lateral giant causes protrusion at the head end. These connexions of the giant fibres to the chaetal muscles are clearly adaptively correlated with their connexions to the sense organs; withdrawal from a painful stimulus at the head end depends upon the tail acting as a holdfast, and vice versa. To be able to withdraw either head or tail requires two through-conducting systems.

If any doubt the effectiveness of these giant fibre responses in relation to the hazards of an earthworm's life, let him go out on a cool wet night with an electric torch and try to catch earthworms as they lie on the grass half out of their burrows.

It is not without interest to reflect upon the similarity between the nervous mechanisms of the sea-anemone and the earthworm. In both there is a slow conducting system, involving cell-to-cell conduction, by which the more leisured activities of the animal are coordinated; in both there is a through-conducting system which comes into action in circumstances threatening the animal's safety. But in detail of course they have less in common. The slow conduction of the anemone (we believe) involves only the synaptic delay in the nerve net, in the earthworm it involves coupled reflex arcs. The through-conduction of the earthworm resides in the giant fibres, whereas the anatomical basis of the anemone's through-conducting system is incompletely known. The same requirement of self-preservation provides the selection pressure, and Nature has evolved two different types of through-conducting system from the two very different types of nervous system she had to work on.

It happens that the annelids are a particularly convenient group for our present purpose and offer an interesting range of central nervous mechanisms. The next animal we shall consider is the leech. Although the large medicinal leech is extinct in Britain, our fresh-

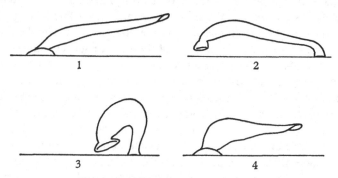

Fig. 41. Looping movements of a leech.

water ponds abound in smaller leeches which are very readily recognized by their looping movements. (Fig. 41). The leeches are more closely related to the earthworms than to the marine worms like *Nereis*, but they are specialized in having a sucker at each end of the body and in having a constant number of segments.

The sequence of the looping movements is as follows.

Stage (*a*). The animal is attached by its posterior sucker, the longitudinal muscles are relaxed (circulars contracted), the body is extended, and the anterior sucker is protruded.

If the anterior sucker comes in contact with a solid object—

Stage (*b*). The anterior sucker takes hold, the posterior sucker lets go, the longitudinal muscles contract (circulars relax), the body shortens, and the posterior sucker is protruded.

Whereas in the earthworm the passage of a wave of contraction is a continuous process and the waves follow one another in smooth succession, the looping of the leech is essentially discontinuous. The sequence can be held up indefinitely in either stage. The coordination of movement in the leech involves not only the stretch-stimulated reflexes of the earthworm, but also the stimulation of tactile sense organs in the suckers.

If the leech is freely suspended in water there is no attempt at looping; instead, the animal usually tries to swim—more of that later. If a cover slip is presented to the posterior sucker it is grasped, the longitudinal muscles relax at once and so on, and the leech remains indefinitely in the stage (a) posture with its anterior end swaying about in the water. If now a second cover slip is presented to the anterior sucker it is grasped and the animal goes into the posture of stage (b). But the transition from one stage to the other is not as simple as the throwing of a switch. The fixation of the anterior sucker is the stimulus which causes the posterior sucker to let go, but it is not the stimulus which causes the protrusion of the posterior sucker; the posterior sucker is protruded only if the body is allowed to shorten. Thus a complex act, whose components appear to be simultaneously executed, proves on analysis to be a sequence of events, the initiation of one being dependent upon the previous completion of another.

We have already seen how tonic reflexes can be coupled up so as to produce harmonious adjustment of the posture of the body. We now have an example of phasic reflexes being coupled up, not in parallel but in series, or, more correctly, in sequence, the consummation of one reflex being the appropriate stimulus for the next. Reflexes thus coupled in sequence are called chain reflexes.

From the looping movements of the leech and the mechanism of their coordination we now turn to the mechanism of the swimming movements. In addition to its longitudinal and circular muscles the leech is provided with dorso-ventral muscles, by whose contraction the body can be flattened out into a sort of ribbon. Thus flattened it is thrown into undulations which pass backwards over it. Swimming in the leech is the same as swimming in the eel or any long-bodied fish except that in the fish the waves are in the horizontal plane whereas in the leech they are in the vertical plane.

From what we know about the passage of peristaltic waves in the earthworm we might be tempted to guess that the swimming movements of the leech could be looked upon as 'peristaltic' waves travelling down the longitudinal musculature, with the dorsal and ventral halves out of phase, that is to say we might seek the explanation in terms of local stretch reflexes and short internuncial neurones. But experiment does not bear this out.

If in Biedermann's experiments on the earthworm we clear away the muscles from the nerve cord for several segments on either side of the cut, the peristaltic wave no longer passes over the region of the injury. Although the 'state of excitation' (one must have some conveniently short expression for the nervous events accompanying the peristaltic wave) can pass some distance along the cord through the internuncial paths it will die out unless reinforced by the cooperation of local reflexes on the way. But if we try the same thing with the swimming leech and clear away all tissues from the nerve cord for a distance of some segments, we find that the swimming movements of the two halves of the body continue to be perfectly coordinated. The state of excitation is thus capable of extending further along the nerve cord in the case of the swimming leech than in the case of the crawling earthworm. And the following experiment shows the same thing even better. The first 8 to 10 segments of the anterior end of a leech are opened and pinned out on a cork. All the nerves between C.N.S. and body wall are cut and electrodes are placed on one nerve for the recording of nerve impulses. The posterior half of the body is allowed to hang into a vessel of water and performs swimming movements. Now although the anterior parts of the body are completely immobilized and although all the reflex arcs are cut, bursts of impulses can be recorded in the nerve. These bursts continue regularly for so long as the posterior part of the body continues to swim, and the timing of the bursts is exactly in accordance with the rhythm of the swimming movements. This suggests the possibility that the whole rhythm is initiated and regulated by the C.N.S. alone. Yet the nerve cord, when completely isolated from the body, shows no activity whatever. The conclusions reached from such experiments are that some kind of nervous rhythm extends over the whole C.N.S. if, but only if, some part of the body is performing normal swimming movements. Although there is no inherent activity in the C.N.S. there is an inherent pattern to which the whole C.N.S. will conform if the activity associated with swimming is initiated in any part of it which retains its normal connexions.

The concept of pattern in the C.N.S. thus emerges from our studies on the leech but is not so well illustrated there as it is in *Nereis*. Although *Nereis* has well developed longitudinal and circular muscles it does not depend upon them for ordinary slow progression. When crawling it uses its parapodia as rather crude limbs which can be bent backwards or forwards, and it distinguishes between the effective stroke and the recovery stroke by protruding the chaetae during the backward movement. The parapodia do not move at random but are coordinated and the coordination problem facing *Nereis* is precisely the same as the coordination problem facing a rowing eight—how to prevent the oars from getting in one another's way. The rowing eight adopt the method of synchronal rhythm, all the oars striking the water together. *Nereis*—and nearly all other animals—adopt the method of metrachronal rhythm; each parapodium is at an earlier phase of its stroke than the parapodium posterior to it and at a later phase of its stroke than the parapodium anterior to it, and the two parapodia of

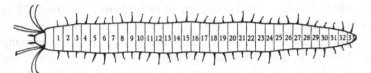

Fig. 42. Diagram to show the disposition of the parapodia
of *Nereis* during crawling.

the same segment are exactly out of phase. To explain this further we must make reference to Fig. 42. In segment 9 the right parapodium is in the middle of its effective (backward) stroke whereas the left parapodium is in the middle of its recovery (forward) stroke. In segment 8 the parapodium on the right is moving backwards but has not reached the middle position; in segment 10 the parapodium has passed the middle position. Thus parapodium 9 is at an earlier phase of its effective stroke than parapodium 10 and at a later phase than parapodium 8.

When *Nereis* is observed in the process of slow crawling the movements of the parapodia convey the impression of waves passing forwards from tail to head. At rest all the parapodia are inclined slightly backwards. When the animal is about to move the first thing that happens is that every fifth (or sixth) parapodium executes a quick forward stroke and the others take up intermediate positions, so that the pattern of the metachronal waves is marked out. This happens almost too quickly for the eye to follow, but with the aid of cinephotography one sees that the development of the pattern starts at the head end and sweeps rapidly backwards. Once it is established over the whole body the waves start to move forwards. Because the pattern assumed by the parapodia of *Nereis* is so obvious, it is easier to accept the idea of central nervous pattern in *Nereis* than in the leech.

We have spent a great deal of time with annelids and on the whole they provide better material for the study of the primitive C.N.S. than do most other phyla. Now let us take a quick look at the two other phyla which have had no mention so far, the arthropods and the molluscs.

If less is said here about the arthropods than has been said about the annelids it is not because less is known—the reverse is true—but because in dealing with the vertebrates and with the annelids we have already brought to light the main features of central nervous activity. The arthropods provide us with further examples of these features, but do not add new features to them. Pattern is well shown in the locomotion of millipedes. These creatures have a large number of segments—about seventy—and, as if seventy pairs of legs was an inadequate number, Nature has provided them with two pairs of legs per segment over most of the body. The legs show a metachronal rhythm of what appears to be the same type as is seen in *Nereis*. There is the same short wave-length—about six segments—the same forward movement of the metachronal waves, the same rapid spread of the pattern when the animal starts to move from rest.

In the lobsters and crayfishes the characteristic escape movement is brought about by a sudden contraction of the powerful flexor muscles of the abdomen which drives the animal backwards through the water. This is mediated by giant fibres. Giant fibres are also present in the cockroach. The hairs on the anal cerci of the cockroach are sensitive to air movements and are in functional connexion with giant fibres which run up through the abdominal nerve cord to the thoracic ganglia. The effect of giant-fibre activity is to set the cockroach running at full speed. The relation between the giant fibres and the neuro-muscular mechanisms of the legs has not yet been worked out, but obviously it must be more complex than the giant-fibre relations with the muscles in the earthworm or crayfish. Biologically, however, the significance of the giant fibres is the same in all—escape. If you try to swat a cockroach and are not very quick about it the air wave moving in advance of the swatter will touch off the cercal hairs; in such a situation the split second difference which giant fibres make is well worth having from the cockroach's point of view.

Reflex arcs of plastic tonus have been demonstrated in the cockroach. The sense organs concerned are campaniform sensilla located in the cuticle of the legs. If a force is applied to the leg such as would be caused by placing a load upon the insect's back, the sense organs increase their frequency of discharge and the faster trains of impulses can be traced through to the appropriate muscles which support the body. Comparing the insect with the vertebrate, let us note that the insect's proprioceptive sense organs are not in the muscle, but in the skeleton, and that, as we saw in an earlier chapter, the processes of facilitation and recruitment go on in the muscle and not in the C.N.S. —yet the two mechanisms achieve the same end, apparently with equal efficiency. Once again in making appropriate use of the materials at her disposal Nature shows herself to be an opportunist. Rather less attention has been given to the molluscan C.N.S. than has been given to the arthropod, which is no doubt partly attributable to the fact that molluscs are not such good experimental material as arthropods. Most particularly does this last apply to the snails and their like, which upon very slight provocation retire into their shells and stay there. Very little can be done in the way of surgery without completely stopping the activities which are the object of study. The majority of lamellibranchs have but one response to any kind of stimulus and that is to shut up and to stay shut up, but here by the removal of one valve of the shell the nervous system becomes accessible in a way that the snail's never does. The cephalopods on the other hand have a higher nervous organization than any other group of invertebrates and have been much studied, but the only feature of their nervous systems we shall consider here is the giant-fibre system, which is extensively developed.

The arrangement of the nervous system in primitive molluscs does not strike one as having been well conceived. Granted that any C.N.S is evolved from some more diffuse system of connexions, it would appear that the molluscan nervous system has been concentrated on three centres where no doubt business was most active. These are the head, the creeping foot and the mantle cavity or breathing chamber which contains the ctenidia. In this way three pairs of ganglia are developed lying far apart in different regions of the body and connected by long nerves (or commissures as nerves are often called when they run from one centre to another). The cerebral ganglia of the head appear to exercise a limited controlling influence over the pedal ganglia and the visceral ganglia. In the lamellibranchs, with the loss of any recognizable head and sense organs, the authority of the cerebral ganglia is still further diminished. In fact the arrangement of the C.N.S. in a lamellibranch such as the fresh-water mussel suggests that the ganglia have simply associated themselves with the three main muscle masses of the body, the anterior adductor, the posterior adductor and the foot (Fig. 43). In the higher molluscs the outlying, semi-autonomous centres are firmly brought under the control of the cerebral ganglia by an evolutionary process which may

be likened to hauling them in by their commissures. In the cephalopods all three pairs of ganglia are fused together in the head.

The most striking features of the cephalopod nervous system are
first the very great development of the sense organs which were dealt
with in the previous chapter, and second the extensive use made of
giant fibres. As in most molluscs the ctenidia of the cephalopod are
enclosed in the mantle cavity, but unlike other molluscs the cephalo-

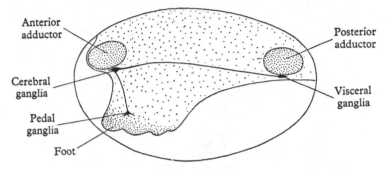

Fig. 43. Diagram of the nervous system of a lamellibranch mollusc.

pods have developed powerful longitudinal and circular muscles in
the mantle wall. Sea water is drawn into the mantle cavity and circulated over the ctenidia in the following way. During inspiration contraction of the longitudinal muscles causes the mantle wall to bulge
outwards; the anterior edge of the mantle is detached from the funnel
and the water enters through the space between them. During
expiration the anterior edge of the mantle engages with the funnel by
a pair of knobs which fit into depressions in the funnel—a sort of
button-hole arrangement—and then contraction of the circular
muscles drives the water out through the funnel (Fig. 44a). The circular muscles are very powerful, enabling the animal to eject water with
great force and drive itself in the opposite direction. The funnel itself
is also muscular and can be directed either forwards or backwards.
This violent ejection of water is brought about through the mediation
of giant fibres and we may note that perhaps this case is an exception
to the general rule that giant fibres are concerned in reactions which
are used only for escape.

The giant-fibre system of the squid is more complicated than any
we have yet considered in that it involves three sets of fibres in series.
The fibres of the first order, of which there are only two, lie within the
brain. The second-order fibres run from the brain to the stellate

ganglia and the third-order fibres from the stellate ganglia to the muscles (Fig. 44b). Each third-order fibre supplies a large group of muscle fibres in the mantle wall. The neuro-muscular junctions are permanently facilitated as in vertebrates, so that the third-order fibre and its associated muscle fibres constitute the motor unit. Another

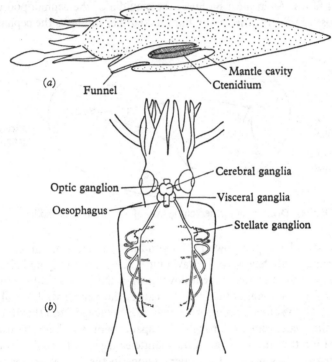

(a)

Funnel

Mantle cavity

Ctenidium

Optic ganglion

Oesophagus

Cerebral ganglia

Visceral ganglia

Stellate ganglion

(b)

Fig. 44. *Loligo.* (a) Median vertical section to show mantle cavity and funnel.
(b) Nervous system (dorsal view).

interesting feature is that among the third-order fibres those which have the longest distance to run have the greatest diameter and hence the greatest conduction velocity. This ensures that contraction is initiated simultaneously throughout the muscle. In Chapter 2 mention was made of the fact that in the mammalian ventricle arrangements are made to ensure simultaneous contraction; the problem here is the same, namely, that if contraction occurred earlier in one part of the muscle the other relaxed parts would be over-distended and possibly ruptured. It has been calculated that the giant-fibre system serves to produce the jet of water in half the time that would be required if ordinary nerve fibres were used.

Now that we have completed this survey of central nervous coordination in invertebrates we may ask whether any further generalizations can be made in the field which we have covered. The first point, which has already been remarked upon, is that the coelenterates and echinoderms have not yielded us any useful information about the beginnings of central nervous function. Our study really began with the annelids, and these we found had many features of central nervous function in common with the vertebrates. We found in the annelids the analogue of Sherrington's spinal arc of plastic tonus, and the term chain reflex which we encountered in connexion with the looping of leeches was first used in connexion with the responses of higher vertebrates. The only major feature not represented in both groups is the giant fibre; and this we can readily understand when we recollect that all the long axons of the vertebrate C.N.S. are myelinated and are capable of conducting impulses even faster than the giant fibres of the earthworm. To judge by the differences in their structure and in their connexions with the rest of the nervous system, giant fibres appear to have been evolved independently in many invertebrate groups and this suggests that natural selection sets a high value upon quickness of response. It also brings home to us that the neurone with the myelinated axon and high velocity of conduction is an evolutionary advance on the cytological level which may have had the most far-reaching consequences for the whole evolutionary history of the vertebrates.

One subject upon which more might be said is the subject of central nervous pattern. In the spinal dog we did not find any evidence of pattern, but pattern is very much in evidence in the C.N.S. of other vertebrates and is particularly well shown in the spinal dogfish. A spinal dogfish has much in common with a decapitated leech. When freely suspended in water both preparations go on swimming indefinitely. But the dogfish has the advantage from the experimenter's point of view that, as in all vertebrates, the sensory and motor nerves separate before entering the spinal cord, so that it is possible to destroy the reflex arcs by cutting the sensory nerve roots without abolishing the contraction of the muscles. In the spinal dogfish, provided that a length of about twenty segments of the body is left intact, the sensory nerves can be cut over the rest and the swimming movements are unimpaired. As with the leech it is essential that one region of the body should be performing normal swimming movements if the C.N.S. is to extend the pattern to the rest of the body.

In our survey we noted pattern in the leech, in *Nereis*, in the millipede, and now we have noted it in the dogfish. We did not observe it in the dog, nor in the earthworm, nor in any of the molluscs we had as examples. Can we say what it is that determines the appearance of pattern in central nervous coordination? The first thing that strikes us is that all those showing pattern are segmented animals and further that as segmented animals they are relatively primitive in having a large number of similar segments. To understand what is involved in being a primitive segmented animal let us recollect the analogy of the rowing eight and the problem of not letting the oars get into one another's way. If you have as many legs as a millipede the only thing you can do with them is to coordinate them into metachronal rhythm, and this pattern of movement entails a corresponding pattern in the coordinating system. The movement of millipedes is a delight to watch but as a means of locomotion it is something to be ashamed of. When one considers the wonderful range of function to which insects have adapted the arthropod jointed appendage one cannot help feeling that millipede evolution has followed unsound lines and that millipedes are not on the list of Nature's triumphs.

Such strictures do not apply to the fishes. It is true that the body is built up of a large number of segments, it is also true that the short-bodied fishes like the whiting are better performers than the long-bodied fishes like the eel, but the fact is that the propulsive mechanism of the fish is very much more efficient than anything which man has produced in his workshops. Here again the problem of coordinating the myotomes is fundamentally the same as the problem of coordinating the millipede's legs. To be brief, wherever we see pattern as a necessary feature of the animal's locomotory mechanism—as is so obvious in primitive segmented animals—we may expect to find a physiological pattern in the activity of the C.N.S.

The reasons why we missed pattern in the dog—for it is there all right—are, first, that the dog, although a segmented animal, has not got a large number of similar segments. But that is only one reason. Had we examined a primitive tetrapod such as the newt we would have seen a very obvious and rigid diagonal pattern in the limb movements. This diagonal pattern can still be seen in the dog when it is walking or running slowly. But if we look at a cat climbing a tree, or better, a monkey, we can realize how the existence of any rigid pattern in the coordination of the limbs would be incompatible with making the best use of the holds which the tree offers. This second reason why we missed pattern in the dog is that in an animal whose limbs have to make adjustments to irregularities in terrain pattern must give way to plasticity. The significant patterns in the dog's C.N.S. are not so much those which relate the position of one limb to the position of the others, but rather those which operate within a single limb, coordinating the sets of muscles which move the joints.

By pattern what we mean is that the motor centres and their associated reflex arcs of plastic tonus are connected together throughout the body and that the position taken up by a limb, or the state of contraction in a set of muscles, is determined more by its central connexions than by its local environmental situation. In the old-fashioned pipe-organ there used to be a stop called 'octave coupler' which, when it was pulled, connected up the keys in octaves, so that if the key of, say, middle C was pressed, all the C keys on the manual went down at the same time. The C.N.S. of *Nereis*, or of a millipede, would appear to contain some device which couples up the appendages not in octaves but in fifths or sixths. Unfortunately neither in *Nereis* nor in the millipede nor in any other animal have histological studies indicated any arrangement of nerve fibres which might form the basis for such a device.

By plasticity we mean the degree to which the central system of connexions can be adjusted to the local situation or over-ridden by influences coming from higher centres, that is, from the brain. This last brings us to the relation between the brain and the coordination of movement. At once it can be said that the so-called brains of the lower animals are not concerned with the coordination of movement; they are concerned in guiding the movement and in deciding whether movement is to occur or not. Worms will crawl and dogfish will swim after the brain has been removed from all nervous connexion with the rest of the body. The same is true for mammals if for 'brain' we read 'the higher centres of the brain'. If the brain stem of a dog or cat is cut in the medulla the animal cannot walk because its organ of balance, the ear, is no longer in connexion with the locomotory mechanism. If the cut is made between forebrain and midbrain the animal can stand up, but goes into a condition known as decerebrate rigidity. If the cerebral cortex—quite the largest part of the brain— is removed the animal can run about quite well. But running doesn't get it anywhere. It continually collides with obstacles. Although its eyes are intact it is unable to react appropriately to what it can see. It can eat if food is placed in its mouth, but not if food is placed in front of its nose. The brain is concerned not in the coordination of movements, but in making the movements appropriate to the environmental situation of which the sense organs make it aware. The more elaborate the sense organs, the greater the range and complexity of the environmental situations which can be recognized and the greater is the task of the brain in analysing this information. For this reason, as a broad generalization, it may be said that the development of the brain reflects the development of the special senses.

10

BEHAVIOUR

Now that we have studied the motor mechanisms of animals, the nature of their sense organs and the part played by the C.N.S. in integrating the sensory and motor systems to maintain posture and carry out simple coordinated movements, we are ready to advance into the wider field of animal behaviour. Of all fields of scientific inquiry none has such a deceptive appearance of simplicity. We see *Paramecium* bump into an obstacle, back away, turn a few degrees and move forward again; we say it is trying to avoid the obstacle. But in fact we have no means of knowing what *Paramecium* has in mind at all. We know why we ourselves do things—at least we think we know and we are prepared to state what we think are our reasons— we go for a walk to get an appetite; we can ask the other man why he goes for a walk, he can tell us and we can believe him if we will. But animals cannot tell us why they do things. To say that an interpretation of behaviour is teleological is to say that it is based upon some premise as to the end or purpose, which is something we have no means of knowing.

Strictly speaking, it is incorrect to say 'the function of haemoglobin is to increase the oxygen-carrying capacity of the blood', for unless you put the haemoglobin in the blood yourself you don't know what its purpose is to do. What you may say is 'the oxygen-carrying capacity of the blood is increased by the presence of haemoglobin'. But very often strict attention to the form of words only leads to pedantic circumlocution about an issue which is perfectly simple. It is easier to say 'the kidney does its best to remove water from the animal as fast as it comes in' than to say 'the output of the kidney is constantly adjusted to the intake of water except in so far as the functional limits of the kidneys are approached', and no one will seriously think that the kidney is being given credit for loyalty and cooperation. This simplified form of expression is called 'teleological shorthand' and is nowadays perfectly respectable. But one must be particularly guarded in using it when matters of behaviour are under discussion, for it is then only too easy to be taken literally.

Any account of animal behaviour which implies motives to the animal is necessarily unscientific. That is not to say that it is wrong

The scientific method has of course its limitations and it is in the field of animal behaviour that these limitations are most severely felt. There are other lines of approach to animal behaviour in which ideas of purpose find a proper place. But here we are concerned with the scientific study of animal behaviour and must therefore avoid introducing concepts which cannot be handled by the methods of science.

Even if one is armed with the foreknowledge of the dangers of an anthropocentric approach, the temptation to put oneself in the animal's place is desperately diffi-cult to overcome. In appraising an environmental situation or in designing the conditions of a laboratory experiment we shall of course have to take into account the capabilities of the animal's sense organs. We may be able to discover that the animal has colour vision. We may be able to deduce from the struc-ture of its eye that it is capable of forming an image of a parti-

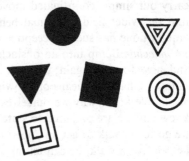

Fig. 45. Six simple geometrical figures. Explanation in the text.

cular object. But this does not mean that the animal makes use of the information which its eye is competent to provide. Nor does it mean that the features which seem significant to us in any situation will be significant to the animal. We know from the number of ommatidia and the visual angle in the bee's eye that it should be possible for the bee to distinguish simple geometrical figures. Six simple geometrical figures are shown in Fig. 45. Nine men out of ten, if presented with these figures and told to arrange them in categories, would arrange them in circles, squares and triangles, because they have done geometry at school and have words for these shapes. Bees on the other hand would show by their behaviour that they put them into two categories, namely, solid figures and concentric figures. What is significant to the bee is the extent to which the figure is broken up into contrasting black and white areas. This reminds us that we have to concern ourselves not only with what the eye can or cannot see but with what the brain does or does not perceive in the situation. The elements of a situation which are significant to (or have valence for) the animal will not necessarily be revealed to us if we try to put our-selves in the animal's place.

As mentioned in an earlier paragraph there are many possible approaches to animal behaviour, but we shall here concern ourselves with only two of these. The first is the *mechanistic* or *causal-analytical* approach, associated with the French philosopher Descartes. He was prepared to consider animals as automatic, thoughtless machines and reasoned that if we fully understood the details of the mechanism we would be able to predict their behaviour in all circumstances. Obviously we are not anywhere near that stage, but the adherents of this view hold that it is possible—at least in theory—to explain all

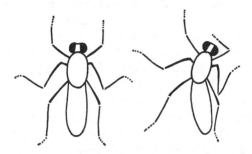

Fig. 46. Effect of brain operation upon the posture of an insect.
(*a*) Normal. (*b*) After removal of left side of brain.

behaviour in terms of the physiological properties of the nervous system. The other approach, the *vitalistic* or *holistic*, starts from the assumption that the animal is more than the sum of its parts and that by taking the animal to pieces, as it were, in the process of analysis we lose sight of those properties of the whole which are the most significant. According to this view a knowledge of nerve physiology, however detailed, will fail to give a complete account of animal behaviour.

Before we go further it might be helpful to enlarge upon these differences in viewpoint with the help of some examples. The best evidence for the mechanistic theory comes from the study of problems such as the following. A certain insect, when placed in a dark room with a single light source, moves towards the light—why does it do so? If we take this insect and remove the left half of the brain we find that after the operation the animal's normal posture is changed. The legs on the left side are well extended, the legs on the right side are drawn up close to the animal's body. This is interpreted as the result of a disturbance in the nervous excitation of the two sides of the animal, expressed in the form of differential tonus in the muscles (Fig. 46). Precisely similar effects can be produced by electrical stimulation of the right-hand side of the brain. When an animal, operated upon in this way, is placed in a room with diffuse light it moves to the right, because the legs on the left side have a mechanical advantage over those on the right. We then take another animal and instead of removing the left half of the brain we cover the left eye with opaque paint. Again the same asymmetrical attitude is taken up in diffuse light and the animal makes circus movements to the right, because the state of excitation in the right-hand side of the brain is greater than in the left. Now consider the normal intact animal in the dark room with the light source initially to its right. The right eye is more strongly illuminated than the left; therefore the legs on the left-hand side are more extended than those on the right; therefore when the animal moves it will circle to the right. It will continue to do so until the intensity of light falling on the left eye is the same as that falling on the right eye. Thereafter it will go straight towards the light. Why does it go towards the light? Not because it likes the light, not because it hopes to find something there but because it just can't help it. Its nervous system is organized in such a way that it cannot possibly do anything else.

Now let us see what the other side have to say to this. They will not at this stage have to defend themselves with the subtleties of philosophical argument. They will say: 'We accept your experiment; we accept its results and its conclusions. But what has it all got to do with animal behaviour? This insect of yours doesn't live in a dark room with a single light source, it lives in open country where it hunts other insects. How is this experiment going to tell us anything about the animal's normal behaviour? You must understand that our observations indicate that an animal appreciates its surroundings not in terms of points of light or darkness but in broader terms of the situation. Let us give you an example. Octopuses eat crabs. If you take a crab on the end of a string and dangle it in the water of an aquarium about six inches above the bottom the octopus shows signs of interest; it will go up to the crab and inspect it and may puff a few jets of water at it. But let the crab drop to the bottom and in a flash the octopus has caught it and started to eat it. A crab dangling on a string in the water is not connected in the mind of the octopus with feeding. A crab running about on the bottom is something quite different; it means food and the octopus reacts accordingly and without delay. In simplifying the situation in order to make analysis easier you remove the essential features. If you want to understand animal behaviour stop your footling experiments, get out into the field and see what the animals really do.'

Faced with these alternatives there can be no doubt about the physiologist's position. On the one hand he is invited to extend his knowledge of the workings of the nervous system to embrace the field of animal behaviour; on the other he is warned off the field and told that his methods offer no hope of useful progress. The behaviour of an animal is the result of the contraction of its muscles which are brought into action by the nerves. No one suggests that there is anything supernatural in these processes or that the physiologist is not competent to investigate them. It should be the physiologist's task to trace the relation between one event and another throughout the nervous system as far as he is able. Eventually it may turn out that there is some paranormal aspect of animal behaviour which defies scientific analysis; but at present there is little point in inventing theoretical difficulties while so many practical problems remain unsolved. In formulating his conceptions of the nervous mechanisms he studies in the laboratory, however, the physiologist will do well to relate his ideas to what the naturalist can tell him of the behaviour of animals in the field.

It is natural, therefore, that we should begin by considering simple manifestations of behaviour if we wish to see how far nerve physiology will take us. We have already had one example, the stimulus in that case being light, and it may be stated at once that the best cases from the point of view of analysis are those in which the stimulus is directional and in which the animal becomes oriented with respect to the stimulus. Responses in which the animal or the track on which it moves is oriented with respect to a directional stimulus are of the type known as taxis. But it is also possible for an animal to find its way from one place to another where the stimulus is not directional and where its movements do not seem to be oriented. This type of response is called kinesis. Kinesis, like taxis, is susceptible to analysis.

Porcellio scaber, a woodlouse, is a crustacean partly adapted to life on land, but having a cuticle which is by no means watertight it has little resistance to desiccation and is normally found in damp places, under logs or stones. It seems to prefer (teleological shorthand) moist air to dry air; at least, if a moist-dry alternative chamber is set up (see Fig. 47) and several woodlice are placed in it, they will later be found gathered together on the moist side. This congregation upon the moist side might be thought to be the result of an avoiding reaction at the moist-dry boundary, such as is shown by *Paramecium* when it encounters an obstacle; but when the animals are observed in movement they appear to cross the boundary in either direction without hesitation. To understand what is happening one has to set up a uniform chamber instead of an alternative chamber and study the animals at a series of different, but uniform, humidities. When this is done it is found that in dry air the animals spend most of their time moving about, whereas in moist air they spend most of their time at rest. What happens in the alternative chamber is that they spend half

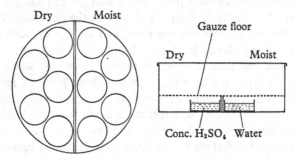

Fig. 47. Alternative chamber. (*a*) From above. (*b*) In section.

their walking time in the dry half and all their resting time in the moist half. Dry air is a stimulus which makes them get up and move and prevents them from resting; this is sufficient to account for the fact that they are found in damp places.

Now it is fairly obvious that we could make a machine that would run about in dry air and come to rest in moist air, and the fact that analysis of this one aspect of woodlouse behaviour has been possible may fairly be quoted in support of Descartes and the mechanists. We may note, however, that much physiological detail is lacking. No one has yet investigated the coordination of locomotory movements in the woodlouse, no one has yet identified the sense organs by which it appreciates differences in humidity. Nevertheless we do know something about coordination in other arthropods and in one arthropod at least a sense organ sensitive to atmospheric humidity has been described. There is nothing about this feature of behaviour which cannot be accounted for in terms of the already known properties of the nervous system. Nor is the physiologist's explanation at variance with the observations of the naturalist.

Taxes and kineses are very simple forms of behaviour. They involve little more than adjustments of the mechanism of posture and loco-motion in response to sensory stimuli of a simple kind. The brain is not required to interpret complex sensory information. The highest forms of animal behaviour are those which we associate with the processes of learning. Between taxes and kineses at the bottom of the scale and learning at the top we recognize that intermediate range of behaviour which we call instinctive.

What we mean by instinct is not clearly defined. Instinctive behaviour is inborn and not acquired by experience—but so also are the ordinary spinal reflexes. To qualify as instinctive, behaviour has to reach a certain standard of complexity. That means we shall see instinct manifested in highly organized animals, capable of complex behaviour. Yet in the highest animals, the mammals, although instinct has its part to play, behaviour is determined more by the animal's own experience. Therefore, going a little lower in the scale, we find our best material in birds and insects.

Instinctive behaviour, then, is inborn and complex. It is also generally adaptive in relation to natural circumstances; it achieves something, like the building of a nest. But it is also true that the animals which rely upon instinct may fail completely in the presence of some unusual element of the situation and are liable to engage in activities which are quite pointless. For example, many ground-nesting birds cover their eggs with vegetation when they leave the nest; if they are chased off before they have time to cover the eggs they still go through all the motions of removing a non-existent covering when they return.

A certain amount of rigidity is thus characteristic of instinctive behaviour. Of this the classical example is the case of Fabre's processional caterpillars. These insects, living a semi-communal life in pine trees, proceed in search of food in single file, head to tail, following a trail of silk laid by the leader. Observing such a procession Fabre induced the leader to march on to the rim of a vase and when it had completed the circuit he broke off the rest of the procession which had not yet reached the rim. The leader, seeing the back end of another caterpillar in front, closed up on it. For several days the procession continued around the rim of the vase; eventually some of the caterpillars got exhausted and fell off.

This is perhaps too simple an example to have chosen. A better one is provided by dung-beetles. These beetles work in pairs, which in itself implies a considerable degree of complexity and also of flexibility in their behaviour. Having found some dung the beetles cooperate to form it into a ball (of diameter about one and a half times the length of the insect); then they roll it away and bury it. In this performance there are some characteristically rigid features. If the beetles are offered a completed ball when they have just started to make one themselves, they ignore it. If, when they have completed their ball are are rolling it away, it is removed and the beetles are replaced on the dung they run aimlessly about and do not start to make a new ball; but now they will accept the ball which they previously ignored. In contrast to this rigidity they show surprising adaptability in overcoming obstacles when they are rolling the ball over uneven ground. They push it or pull it, or get underneath and heave it up. One observer fixed the ball to the ground with a long thin stake; the beetles reacted to this by cutting the ball in half, freeing it from the stake and joining the halves together again. This looks very like the intelligent appreciation of a problem, but on further investigation it turns out that cutting the ball in half is a purely instinctive act which is carried out when all other attempts to move the ball have failed.

It is true that to the nerve physiologist there is much in instinctive behaviour which is suggestive of chain reflexes. Instinctive acts of behaviour are, as we have seen, complex and made up of sequences of component acts; very often the successful completion of one component act is necessary for the initiation of the next. If the rook drops the twig which it is carrying back to build its nest it does not go down and pick it up; it goes on to the nest, settles and then takes off again and fetches another twig. But no physiologist has given a convincing analysis of a complex instinctive act in terms of chain reflexes, no doubt largely because the stimuli initiating such responses are so complex. Game cocks will instinctively attack one another on sight; but it is difficult for the physiologist to picture for himself in terms of nerve impulses how the image of a cock bird upon the retina of another cock has certain significant properties not shared by the image of a hen. Superficial talk of chain reflexes helps very little and naturalists studying instinctive behaviour have had to develop their own systems of thought and methods of interpretation.

Learning has been described as an adaptive change in individual behaviour resulting from experience. Most forms of learning are associative. But we know that some animals are able to memorize the details of their surroundings although we are not able to show that they do so in association with particular acts; this is called *latent learning*.

The form of associative learning which is best understood is the formation of a *conditioned reflex*, a term which we owe to Pavlov. Pavlov was originally interested in digestion and in the control of the secretion of the digestive juices. He was engaged in studying the reflexes of the salivary glands and found that his experiments were being upset by the ability of his dogs to anticipate the appearance of the food from various preparatory moves in the experiment. This led him to see in the salivary reflex a means of studying the higher functions of the brain. When food is placed in a dog's mouth there is a flow of saliva; this is an ordinary reflex and all dogs show it. If a bell is sounded regularly just before feeding time the animal will come eventually to secrete saliva at the sound of the bell, before it has received the food. The salivary response to the ringing of the bell is a reflex which has been acquired as a result of the conditions of the dog's existence. An acquired, or conditioned, reflex can only be formed upon the basis of an inborn, or unconditioned, reflex; but there is no limitation upon the sort of stimulus to which a conditioned response can be developed. The properties of conditioned reflexes in mammals have been worked out in great detail, and it is also important to know that conditioned reflexes are not confined to mammals. They have been established in other vertebrates, in insects, in molluscs; even earthworms, creeping through tubes, can learn always to turn to the right at a junction if they habitually suffer an electric shock after a short progress to the left. The training experiments which were described in Chapter 8 are of course further examples of the formation of conditioned reflexes.

Habituation is the formation of a conditioned reflex in the negative sense. It consists in learning not to respond to irrelevant stimuli in the way that a gun dog learns not to be scared by the sound of the gun.

Trial and error learning is another form of associative learning. It has been much studied in rats, a favourite design of experiment being to require the rat to find its way through a simple maze in order to reach its food. After several repetitions of the experiment the rat learns to avoid the wrong turnings and can run straight through the maze to its goal. The essential difference between trial and error learning and the formation of a conditioned reflex is that in the latter case the initiative is with the experimenter (or with nature) who applies the conditioned stimulus at the same time as, or just before, the unconditioned stimulus, whereas in trial and error learning the initiative is with the animal. While the ability to form conditioned reflexes is widespread among animals, the ability to learn by trial and error is more or less confined to inquisitive—pecking, sniffing— sorts of animals, such as rats, and mice and small birds. Animals of a placid or lethargic disposition, like cows or frogs, do badly.

Insight is the appreciation of the relations of a situation and the sudden adaptive reorganization of behaviour, not preceded by trial and error. As usual, an example gives a better explanation than a definition. If a hen is confronted with a dish of food upon the far side of a wire-netting fence—say, ten yards long—it will flap unavailingly at the fence until it gets tired. A dog in the same situation will run round the end of the fence and get the food at once. The dog has insight, it perceives the relations of the situation; the hen does not. Insight is hardly at all developed in animals other than the mammals.

When we say that insight does not involve trial and error all we mean is that the animal does not attempt to put incorrect solutions into practice. Objectively we have no means of knowing how the animal arrives at the right solution. In ourselves, however, we know subjectively that in working out problems in our heads we often proceed by trial and error. In simple trial and error learning—the rat in the maze—the animal becomes aware of its errors by its failures in practice. What we seem to do is to compare the trial solution with the background of our experience and accept or reject it according to its compatibility or otherwise. Making full allowance for the dangers of trying to put oneself in the animal's place it does not seem unreasonable to suggest that there is a possibility—no more—that insight is a form of trial and error learning in which associations are made between a present situation and events which are remembered from the past.

The possibility that insight is a form of trial and error learning is of some interest because a modern computer can find solutions to numerical problems by trial and error. It can be so designed that, once a solution has been reached by trial and error, its operation in dealing with subsequent problems is modified according to the solution which has been reached; that is to say, a machine can modify its behaviour as a result of experience, it can learn. A computer can memorize information. It would be possible to programme a computer to solve a problem by trial and error, by a process involving the use of memorized information, and then to embark upon some further action after the solution had been found; such a machine would show insight in the sense that was suggested in the previous paragraph. All this may have little relevance to the physiological study of animal behaviour; but on the other hand it shows that no one is in a position to say that there is any well-authenticated feature of learning which cannot in principle be shown by a machine.

There are difficulties, however, in the way of regarding the brain as a computer in any way comparable with man-made computers. We know next to nothing about the workings of the brain in any animals other than mammals, and among mammals more is known about the human brain than about any other. Not unnaturally investigators have been concerned with localization of function in the brain, making special study of the impairment of function resulting from brain wounds and tumours in man and making operations upon the brains of other primates. The occipital region of the cerebral cortex is found to be concerned with vision, the temporal lobes with hearing and speech and so on. Now central nervous tissue in mammals has no power of regeneration, so that if any part of the brain is removed it cannot be replaced. In spite of this there is usually very considerable recovery of the function which is initially lost as the result of an operation. In rats it has been found that more than half of the cortex can be removed without permanently abolishing conditioned reflexes previously acquired. This means that although there is some vague association between certain regions and certain functions in the brain no one part of it is exclusively concerned with some function which cannot be carried out by another part. It is very difficult to see how this can be the case if the operation of the brain depends upon the way in which it is 'wired up', as does the operation of a computer. Simple analogies with telephone exchanges and computers will not help us to understand the way in which the brain works.

We have now reached a stage at which it would be profitable to return for further comment to some of the ideas about animal behaviour which were put forward in the earlier pages of this chapter. We set out to bring animal behaviour into line with nerve physiology. Very often, as with simple taxes and kineses, our attempts at analysis have been reasonably successful. We have reason to believe that in many cases it would be possible to make a machine which would behave in the same way as the animal. But if we were required to build such a machine we would want complete freedom in design and in choice of materials, and we would almost certainly want to make extensive use of such devices as valves, relays and the like. It would nevertheless be fair to ask whether we could design a machine using nature's materials and methods—although of course we could not be expected to build it. And here we find ourselves rather at a loss. We have a very good idea of the way in which the peripheral nervous system works, but as we saw in the last chapter we know very little about what goes on in the central nervous system. While this gap remains in our knowledge we can hardly make any serious attempt to follow in nature's footsteps. Our survey has shown the possibility of analysis and interpretation of behaviour on mechanistic lines but we must admit that our knowledge of nerve physiology has helped us very little.

The difficulties confronting the physiologist have already been stated but it will do no harm to state them again. First, there is the purely technical difficulty of applying to the central nervous system those methods which have seen so successful with the peripheral nervous system. Second, there is the difficulty that even if we were able to follow nerve impulses through the C.N.S. the complexity of all but the simplest central nervous processes would pass beyond the limits of comprehension. A similar situation arises in the physical sciences. The physical chemist cannot keep track of the action of each and every molecule of ethyl acetate undergoing hydrolysis in a test-tube; but he can say with some accuracy what proportion of the total number of molecules will have broken down after a certain time. In the same way the physiologist will have to give up chasing individual nerve impulses; but as yet he has not found any simple system of formulating his ideas about central nervous phenomena which is at the same time fully compatible with what is known at the more detailed level.

Hitherto we have concentrated upon the simplest examples of animal behaviour, even when we had to consider its higher manifestations as in learning, and this may perhaps have given a false impression of simplicity. The next example is described for no other purpose than to supply a suitable corrective.

Bee-keepers and those who have made a special study of bees have long been aware that the bees tend to concentrate their activities in places where nectar is abundant, and have suspected that the bees had some means of communication. The investigation of this problem was undertaken by von Frisch in the early 1920's. His first experiments confirmed the general impression. He caught some bees, fed them on sugar solution and liberated them at a certain station; presently a large number of other bees would be seen buzzing around. The concentration of the bees was much more obvious if scent was introduced. In later experiments he allowed the captured bees to feed from a small dish of sugar solution which stood upon a filter paper impregnated with the scent of some flower. Before the bees were released the sugar solution was removed and other filter papers impregnated with other scents were set out. When the bee crowd arrived they made straight for the scent upon which the captured bees had been fed. Further, the bee has itself a scent organ which it can use to mark the place at which it has fed. Scent thus played an important part in communication, but was not the whole story.

Von Frisch then went in for marking bees with a small splash of paint, and he prepared an observation hive with glass slides in which he could observe their behaviour. Bees were captured, marked and

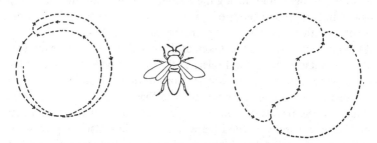

Fig. 48. Dances of the honey bee, with insect drawn to scale.
(a) Round dance. (b) Waggle dance.

fed, and released at a certain place, while at the observation hive von Frisch waited to see what the marked bees would do. As soon as they came in they made their appearance upon the vertical surfaces of the combs and ran about in seemingly aimless circles. On closer observation, however, a certain regularity was apparent. The bee, newly arrived, would make one or two complete turns to the right, then reverse and make one or two complete turns to the left; and other bees, unmarked, working on the comb, would join in and imitate these movements. This particular manoeuvre von Frisch calls the *Rundtanz*, or 'round dance' (Fig. 48 a). But at the same time he noticed other bees, not marked, which had apparently been collecting pollen, behaving in a rather different way. They would take a few steps forward, the abdomen being swung rather emphatically from side to side during this movement, then circle round quickly, either to right or to left, and repeat. This he called the *Schwänzeltanz*, which in English translations is generally rendered as the 'waggle dance' (Fig. 48 b).

Many other observations and experiments, which would take too long to describe, led von Frisch to conclude that the round dance was a signal which meant that a source of nectar had been discovered and would be found in association with the scent adhering to the dancer, and that the waggle dance similarly indicated the discovery of a supply of pollen. This was published in 1923. Although the main feature of the discovery, i.e. that the dances are a means of communication, is still true, the exact significance attached to the dances by von Frisch at that time turns out to be incorrect. The real story is very much more exciting.

Various observations which he made from time to time during later years suggested to von Frisch that there was more to it than he had first thought, and during the Second War he took the matter up again. In this second series of experiments he made use of a number of observation stations set up in the neighbourhood of the hive. Simultaneous observations taken at these stations soon showed that after bees had been fed and released the new arrivals gathered at the place of release or at the observation stations close by; the other stations were more or less deserted. von Frisch therefore had to conclude that the bees were able not only to communicate to each other the fact of a supply of food being available, but also to give some indication of its position. Now in order to establish this point he had been working with stations rather further from the hive than in his earlier experiments; and when he went on to observe the behaviour of marked bees in the hive he found that they were doing waggle dances, not round dances. He therefore made a careful check upon his original work, setting out two feeding stations, one at 12 m and the other at 280 m from the hive, and marking the bees with two different colours so that there should be no mistake about the station at which they had fed. He then observed that the bees feeding at the 12-m station did round dances, those feeding at the 280-m station did waggle dances. Given this clue that the type of dance was related to distance from the hive, he was not long in establishing that the changeover took place somewhere between 50 and 100 m.

This, however, was not the whole answer. The experiments had already shown that the bees must be able to communicate distance more exactly than just nearer or further than 75 m. In his next series of experiments he observed and compared the behaviour of bees coming from different distances and was able to discover that the character of the waggle dance varied with distance. Bees coming from short distances made short and frequent runs, those coming from further away made longer runs, but less frequently.

Here, then, is the way in which bees communicate distance—by the frequency of the waggle runs they make. How do they communicate direction? Continued observation established the following facts.

Number of waggle runs in 15 s	Distance (m)
10	100
5	700
2·5	2,500

First, marked bees from a single station all make their waggle runs in the same direction; other bees make waggle runs in all other directions. Second, marked bees from two stations in opposite directions from the hive do their waggle runs in opposite directions; this gives the clue that the direction of the waggle run is significant. Third, with marked bees all from a single station the direction of the waggle run is progressively altered throughout the day; this gives the clue that the direction of the sun is probably sigificant. The pieces of the puzzle then fell very quickly into place. Vertically upwards on the surface of the comb indicates the direction of the sun; right or left angles from the vertical mean right or left angles from the direction of the sun. Thus in Fig. 48b the message is: go outside the hive, face the sun, turn 30° to the right.

The story of the round dance is still the same; it means nectar to be found near the hive in association with the scent adhering to the dancer. It is a message to the close-foraging workers. It is perhaps useful to mention here that worker bees graduate from tending the young to building combs, to foraging near the hive and finally to foraging at a distance. The waggle dance is a message to the distant-foraging workers. It gives range with an average error of 100 m and bearing with an average error of 3°.

There is a great deal more to this tale and all of it is of fascinating interest—how, when the food lies beyond an obstacle the bees give the bearing across the chord and the range round the arc, how they are able to appreciate the position of the sun when it is behind a cloud by the polarization of the light from a patch of blue sky. But these are the frills. What is staggering about the business is the elaborate encoding and decoding of sensory impressions. The distance flown has to be translated into the frequency with which waggle dances are performed. The direction of gravity is used to represent the direction of the sun. How has all this been evolved? It may be that we make rather heavy weather of things like angles because we have had to learn so much about their properties at school; it may be that the idea of an angle comes quite easily to the bee's brain. Yet making all the allowance one can for the special conditioning of the human brain, the whole thing seems quite a formidable feat for an animal whose brain is about 2 mm across. Whether it turns out to be a part of the bee's instinct or whether it is something that the bee learns by watching others, the fact remains that as yet we cannot match it in any other animal. Clearly, we shall have to readjust our ideas about the functional complexity that can be attained with a limited number of nerve cells. And if perhaps we have been encouraged by the promise of some earlier attempts to analyse animal behaviour, we should be sobered at the thought of the tasks that lie ahead, even among the invertebrates.

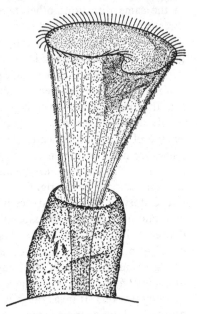

Fig. 49. *Stentor*.

Here is one last problem. *Stentor* is a ciliate protozoon, a relative of *Paramecium*. It is a funnel-shaped creature, attached by its base to a short tube and using its cilia to create a current of water from which it extracts small particles of food (Fig. 49). If a stream of fine carmine particles is allowed to fall upon it there is at first no response. Then it starts bending its body, first to one side, then to the other. If the particles still fall upon it, it adopts different methods; the cilia stop and momentarily reverse the direction of their beat. If this fails to give relief the animal will withdraw into its tube for a few moments and then emerge again. This will be repeated a few times. Finally if the nuisance continues it looses its hold on the tube and swims away. Here we have an example of behaviour made up of a series of acts or phases, each act simple yet coordinated; the whole behaviour is highly flexible and unmistakably adaptive to the circumstances. It is on about the level that we should expect from a coelenterate or a platyhelminth. But this is a protozoon, an animal without any nervous system at all!

APPENDIX I

A GUIDE TO FURTHER READING

I shall assume that you are a student and that you are already in (or about to begin) your first year at university; that in biology you have been introduced to cell biology, to structure and function in the mammalian body, and to the main phyla of the animal kingdom; that you have devoted about as much time to chemistry and physics as you have to biology; and that you now want to know something more about physiology.

Let us suppose that what you want to know is just something more about some particular fact which received brief passing mention in this book—say it was echo-location in bats. What you do is to go to the library and get hold of an advanced text-book of physiology (not one entitled human physiology for doctors aren't interested in bats); *Comparative Animal Physiology* by Prosser and Brown would be an obvious choice. On looking up bats in the index you pick up a page reference to acoustic orientation. Under this heading there is about one and a half pages of text. If this is not enough for you, there are nine references given in the text; and looking them up at the end of the chapter you find that one of them is a book of 413 pages by D. R. Griffin called *Listening in the Dark.*

But it may be that you want to know more about physiology as a subject. Perhaps you think that this is a subject you would like to do research in some day. In that case you need guidance of a different sort.

It is convenient to recognize three divisions of the subject: mammalian physiology, comparative physiology and general physiology. Mammalian physiology is in a special position because of its importance for medicine. This has meant that a vast amount of research effort has been applied to it and that in consequence we know far more about the physiology of the mammal than about the physiology of any other class of animal. Comparative physiology seeks to extend this corpus of knowledge to include animals of other classes. Because the pageant of the animals is so vast and varied, and because the number of research workers is relatively so few, comparative physiology has not been penetrated so deeply. In surveying the various manifestations of a physiological function over a wide range of animals the comparative physiologist is also seeking generalizations which will enable him to identify the fundamental problems of general physiology. In general physiology we are concerned with the interpretation of physiological function in terms of physics and chemistry.

All these three branches of physiology are becoming increasingly closely interwoven. Most of our knowledge of general physiology has come from the study of mammals, but not all of it. The decisive advance in the general physiology of the nervous system, which threw a flood of light upon mammalian physiology, was made on cephalopod molluscs. It is a further, practical aim of the comparative physiologist to discover new situations wherein the biophysicist and the biochemist can deploy their methods to best advantage, situations which must be sought mainly among the invertebrates—a rich field whose exploration has barely begun.

I am now going to suppose that your interest is in the physiology of the lower animals, that is, in comparative physiology and possibly in general physiology.

A sound though not necessarily detailed knowledge of mammalian physiology is a good foundation for both comparative and general physiology. What I have in mind corresponds to the minimum professional requirement for the medical student. Many excellent textbooks of about this standard exist and some of the better known have been kept up to date in a series of new editions. In this category I place *Human Physiology* by F. R. Winton and L. E. Bayliss (5th edn., 1962, 621 pp., Churchill). This book takes the subject at a reasonable pace and is eminently readable. I do not suggest that the first thing you must do is to sit down and read it from cover to cover. But if you want to know more about mammalian physiology dip into this sort of book. The larger text-books of the order of 1500 pages or more are to be avoided except as books of reference.

Comparative physiology being a relatively new subject, fewer textbooks are available to choose from. A good book to begin on, by reading those chapters relating to your special interests, is *General and Comparative Physiology* by W. S. Hoar (1966, 788 pp., Prentice-Hall). This is addressed to the zoologist and does not make great demands upon his knowledge of physics and chemistry. A little more exacting in this respect is *An Introduction to General and Comparative Physiology* by E. Florey (1966, 689 pp., W. B. Saunders). *Comparative Animal Physiology* by C. L. Prosser and F. A. Brown (2nd edn., 1961, 661 pp., W. B. Saunders) is, as already mentioned, a reference book. It is an advanced student's guide to the literature, not a beginner's introduction to the subject.

One can soon learn to recognize the category to which a book belongs merely by turning over its pages. The more advanced the book the more the text is interspersed with references. In an advanced book the illustrations are usually taken unmodified from original papers, with acknowledgement, whereas the illustrations of an elementary book are more often designed by the author to explain concepts rather than to display the results of an experiment.

On reading in general physiology it is less easy to give advice. A good book to begin on used to be *General Physiology* by B. T. Scheer (1953, 512 pp., John Wiley) but this is now becoming out of date. *A Text Book of General Physiology* by H. Davson (3rd edn., 1964, 1135 pp., Churchill) is excellent, up to date, but heavy going for the beginner.

The difficulty about general physiology is that one cannot get very far with it before the need for a good background in physics and chemistry begins to make itself felt. There is a real problem here, for if a university student is at the stage before specialization begins in earnest and is studying biology, physics and chemistry together he will find that the sort of physics and chemistry he is being taught within those disciplines is not the sort of physics and chemistry he needs for biology. What the biologist needs to know about are membrane phenomena and diffusion, things which are of secondary interest to physicists and chemists. The general physiologist usually has to learn his physics and chemistry by working on his own, though of course the more he already knows of the general aspects of these subjects the easier it is for him to make progress.

Some books on general physiology quite openly assume that the reader is well versed in the physical sciences. Those that do not assume this take refuge in the didactic method, presenting mathematical formulae without indicating how they are derived. One author, however, has set himself the task of rescuing the biologist who has parted company with physics and chemistry when he left school and is now calling for help: W. M. Clarke has written an attractive and readable book called *Topics in Physical Chemistry* (2nd edn., 1952, 731 pp., Williams and Wilkins) for the benefit of the medical student who is likely to go into research. It is in no sense a medically biassed book and any biologist who feels the need of support can read it with pleasure as well as with profit. Another book for the would-be biophysicist, of specialized content but beyond all praise, is *Nerve, Muscle and Synapse* by B. Katz (1966, 184 pp., McGraw-Hill). We could do with a lot more books like these.

Behaviour is generally regarded as coming under psychology rather than physiology, but since I have included in this book a chapter on behaviour I ought to say something about it here. If you are interested in this subject I suggest that you take a look at *Animal Behaviour* by R. A. Hinde (1966, 446 pp., McGraw-Hill).

Altogether this is a formidable list, and when you feel some of these books in your hands their very weight is a depressing reminder of how much there is to be known. But you are in no hurry. At your present stage you don't want to cram yourself full of facts. Go to the library and spend an hour or two just looking around. Read a chapter here and a chapter there. Pick on some point that interests you and pursue it through the literature just to see where it takes you. In this way you will begin to get the feel of the subject; you will get a few glimpses of it as it appears from the research worker's bench, see its problems and guess at the demands it is likely to make on you. I don't think you will find this dull.

CLASSIFICATION OF THE ANIMAL KINGDOM

This brief classification is provided for the convenience of the reader who wishes to refresh his memory. It is intended to show the systematic positions of animals which have been mentioned in the text by their popular or generic names.

PROTOZOA: unicellular animals; *Paramecium, Stentor.*
COELENTERATA: sea-anemones, jellyfishes; *Hydra.*
PLATYHELMINTHES: flat worms; *Dendrocoelum.*
NEMATODA: round worms.
ANNELIDA: ragworms, earthworms, leeches; *Nereis.*
ARTHROPODA—
    CRUSTACEA: crabs, lobsters; *Astacus, Carcinus, Maia, Porcellio.*
    ARACHNIDA: spiders.
    INSECTA: insects; *Aedes, Drosophila, Dytiscus.*
MOLLUSCA—
    GASTEROPODA: snails, limpets; *Helix, Planorbis.*
    LAMELLIBRANCHIATA: mussels, clams; *Mya, Pecten.*
    CEPHALOPODA: octopuses, squids; *Loligo.*
ECHINODERMATA: starfishes, sea urchins.
VERTEBRATA: fishes, amphibians, reptiles, birds and mammals.

INDEX

adaptation, sensory, 87
all-or-none law, 78
ATP, 33–5

balance and hearing, 92–6
bees, dances, 140–4
behaviour, mechanist-vitalist controversy, 131–3

chemoreceptor, 89–90
coelom and haemocoel, 25–7
conduction, nervous
    decremental, 107
    through-conduction, 107–10

diffusion, 18–19, 32, 35, 36–7
digestion
    external, 9
    extracellular, 9–10
    intracellular, 5–6
    microorganisms in, 8–9

endocrine, see Chapter 5
energy, 1–2, 33
enzyme, 3, 13–14
excretion, nitrogenous, 48–9

facilitation, 82–4, 109

gastro-vascular system, 22

haemocoel and coelom, 25–7
haemocyanin, 39–41
haemoglobin, 38–9
hearing and balance, 92–6
heart
    control, 30–1
    two-chambered, 20–1
homeostasis, 54–9
hormone, see Chapter 5

inhibition, nervous, 105, 110–11
instinct, 135–6

kinesis and taxis, 133–5

learning, 136–8

mechanoreceptor, 90–6
muscle
    mechanism of contraction, 70–1, 76
    speed of contraction, 71–2
    striated and unstriated, 69–70

nerve fibre, giant, 115–16, 121, 123–4
nerve impulse
    electrical nature, 79–81
    frequency, 82
    velocity of conduction, 80
neurone, structure, 77–8
neurosecretion, 64–5

oxygen
    consumption, 46
    dissociation curve, 39–41
oxygen capacity, blood, 39

pattern, CNS, 119–21, 125–7
peristalsis, 19–21, 112–15
photoreceptor, 96–100
pressure, blood, 17–18, 29–30

reflex, 104–6, 121
    chain reflex, 117–18
refractory period, 79
regulation
    ionic, 49–50, 54–7
    osmotic, 50, 54–7
rhythm, metachronal, 120–1

skeleton
    compliant-elastic, 74–5
    hydrostatic, 73
    jointed, 73–4
sense organs, classification, 86–7
synthesis, 3–5

taxis and kinesis, 133–5
tracheal system, 42–5

urine formation, 51–3

vitamin, 3–4